NONLINEAR ASTROPHYSICAL
FLUID DYNAMICS

ANNALS OF THE NEW YORK ACADEMY OF SCIENCES
Volume 617

NONLINEAR ASTROPHYSICAL FLUID DYNAMICS

Edited by J. R. Buchler and S. T. Gottesman

The New York Academy of Sciences
New York, New York
1990

Library of Congress Cataloging-in-Publication Data

Nonlinear astrophysical fluid dynamics / edited by J.R. Buchler and S.T. Gottesman.
    p.   cm. — (Annals of the New York Academy of Sciences, ISSN ISSN 0077-8923 ; v. 617)
    "Papers in this volume were presented at the Fifth Florida Workshop in Nonlinear Astronomy . . . which was held on October 5–7, 1989, in Gainesville, Florida"—P.
    Includes bibliographical references and index.
    ISBN 0-89766-629-1 (cloth : alk. paper). — ISBN 0-89766-630-5 (paper : alk. paper)
    1. Fluid dynamics—Congresses.   2. Astrophysics—Congresses.
3. Nonlinear theories—Congresses.   4. Astrometry—Congresses.
I. Buchler, J. R. (J. Robert)   II. Gottesman, S. T. (Stephen T.)
III. Series.
Q11.N5   vol. 617
[QB466.F58]
523.01—dc20
                                        90-28426
                                          CIP

SP
Printed in the United States of America
ISBN 0-89766-629-1 (cloth)
ISBN 0-89766-630-5 (paper)
ISSN 0077-8923

ANNALS OF THE NEW YORK ACADEMY OF SCIENCES

Volume 617
December 31, 1990

NONLINEAR ASTROPHYSICAL FLUID DYNAMICS[a]

Editors
J. R. BUCHLER and S. T. GOTTESMAN

Conference Organizers
J. R. BUCHLER, G. CONTOPOULOS, S. T. GOTTESMAN, J. H. HUNTER

CONTENTS

[a]The papers in this volume were presented at the Fifth Florida Workshop in Nonlinear Astronomy, entitled Nonlinear Astrophysical Fluid Dynamics, which was held on October 5–7, 1989, in Gainesville, Florida.

Financial assistance was received from:
- COLLEGE OF LIBERAL ARTS & SCIENCES, UNIVERSITY OF FLORIDA

Preface

J. ROBERT BUCHLER[a] AND STEPHEN T. GOTTESMAN[b]

[a]Department of Physics
University of Florida
Gainesville, Florida 32611

[b]Department of Astronomy
University of Florida
Gainesville, Florida 32611

This Florida Workshop was the fifth in a series of workshops in nonlinear astronomy whose proceedings have been published in the *Annals* of the New York Academy of Sciences. It brought together a group of active astrophysicists with a broad experience in astrophysical fluid dynamics, both analytical and numerical. The workshop, a collaborative effort between the Astronomy and Physics Departments, was organized by J. R. Buchler, G. Contopoulos, J. H. Hunter, and S. T. Gottesman.

We wish to express our thanks to Prof. W. Harrison, Dean of the College of Liberal Arts and Sciences, for the financial support that made this workshop possible. We are also greatly indebted to Dr. Michael Norman from the National Center for Supercomputing Applications at the University of Illinois for organizing the concomitant video publication and to Prof. Pierre Ramond, Director of the Institute for Fundamental Physics at the University of Florida, for providing additional financial support for it.

The Hot Bubble and Supernova Calculations[a]

STIRLING A. COLGATE

Los Alamos National Laboratory
Los Alamos, New Mexico 87545

The recent calculations of James Wilson and Ronald Mayle[1] showed that the mechanism of supernova explosions caused by collapse to a neutron star now appears to be both understood conceptually and modeled convincingly up to 3.4 s following collapse. In particular they show the formation of a hot, high-entropy bubble (FIG. 1) that continues to push on the shocked matter for a long time, enough that the subsequent history is not in significant doubt. The hot bubble is formed primarily due to mu, tau neutrino antineutrino annihilation as first proposed by Goodman, Dar, and Nussinov.[2] The hot bubble that separates the neutron star from the ejected matter has high entropy, 10^2 to 10^4, measured in units of the Boltzmann constant, k, per free nucleon. This high entropy means that for every nucleon there are many photons and electrons (pairs), and so the molecular weight is small and the scale height is large even at modest temperatures, ≤ 1 MeV. Such a photon gas can "push" simultaneously on both the neutron star surface as well as on the expanding matter. It extends to a radius of 10^9 cm, so that no fallback or reimplosion of any significant fraction of the ejected matter will take place. The kinetic and internal energy minus the gravitational energy of matter, whose total energy is positive, is 0.35×10^{51} ergs at 3.4 s. Wilson and Mayle expect this to increase to 1 to 1.5×10^{51} ergs by the end of the calculation, typical of Type II supernovas.

It has long been my major concern (Colgate[3]) that despite a very strong shock wave, a significant fraction of the ejected matter would subsequently fall back onto the neutron star. This fraction, several solar masses or more, would fall back in a time of about a half an hour despite an initial large *positive* velocity (greater than escape) *but ultimately* leading to a black hole.

The reason for the large fallback mass is twofold. First, normal matter falling onto a neutron star will be cooled by neutrino emission in a few tens of seconds from the high temperature, 1 to 2 MeV, that is created when this matter is shocked and compressed at the neutron star surface. Hence, the pressure that might ordinarily extend from the neutron star surface to the shocked matter, forming a piston to maintain the shock, disappears. The neutron star then acts like a black hole. Second, the internal energy in shocked material is always equal to the kinetic energy of motion of the same matter (behind a strong shock). Hence, despite the radially outward velocity that exists behind a strong ejection shock, there is always enough heat to allow for an expansion inwards or backwards (in one dimension) at a velocity equal to the shock-created outward velocity. The radially inward velocity of the rarefaction wave overcomes the original outward radial velocity of the shocked

[a] This work was performed under the auspices of the U.S. Department of Energy.

1

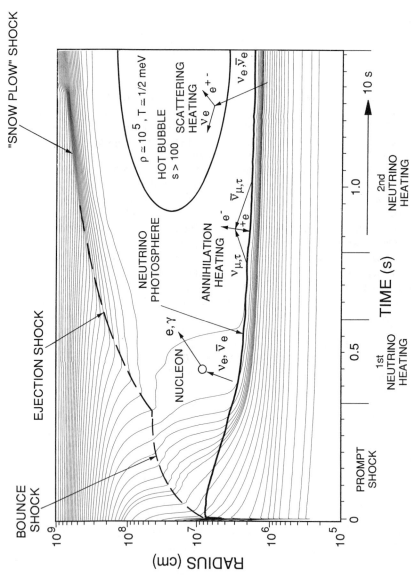

FIGURE 1. Collapse to a neutron star of a Type II supernova according to Wilson and Mayle.[1] This initial bounce shock is relatively weak, but is soon ≃ 0.5 s, strongly reinforced by electron neutrino heating. Inside the ejection shock a hot bubble of high entropy is formed due to $\nu_{\mu,\tau} - \bar{\nu}_{\mu,\tau}$ annihilation. This hot bubble ensures that the matter behind the ejection shock does not subsequently fall back onto the neutron star with subsequent collapse to a black hole.

matter. When one adds to this backwards expansion the strong gravity existing in the central regions, the rarefaction velocity is quite sufficient to ensure reimplosion, even for the most energetic shocks whose total energy greatly exceeds that initially needed to eject matter from the star. Third, if the pressure deficit at the neutron star surface, due to neutrino cooling, occurs significantly later than the formation of the strong outward-going shock, then the resulting rarefaction wave moving in the co-moving matter will always catch up to the shock wave and weaken it. (The flow behind a shock wave is always subsonic.) Thus, we require a positive pressure extending from the neutron star surface to the shock until the shock reaches the outer low-density layers of the star where the entropy generated by the shock alone is very large.

With spherical convergence the problem is more complicated due to what is known as Bondi accretion. Here if the shocked matter has a very high entropy or sound speed, the accretion or rarefaction flow tends to "choke" at a smaller radius corresponding to where the gravitational potential is equal to the specific internal energy (sound speed^2) of the shocked matter. For example, if the shock wave were four times stronger, that is, has four times more energy per mass than is required to unbind the matter, then the radius at which the accretion would be choked by Bondi accretion (Bondi[4]) is no smaller than half the radius of the shock wave, and hence, would only reduce the flow by a factor of between 2 to 4. Thus, there has always been the need to prevent the fallback of matter after being shock ejected from the neutron star.

The reason that measuring entropy in photons per nucleon is conceptually convenient is the following: the gravitational binding of a nucleon to the neutron star is approximately 1/10 its rest mass or 100 MeV per nucleon. (The neutron star cools in several seconds such that its radius is $\simeq 10^6$ cm.) Consequently, at temperatures typical of the neutron star surface, namely an MeV or 10^{10} degrees, the scale height of normal matter, that is, just nucleons, would be roughly 1/100 of the neutron star radius. And so, despite the high temperature, nucleonic matter would be tightly bound to the neutron star, and the "push" 100 scale heights away at the shock contact surface would be trivially small (e^{-100}). Hence, one requires an extremely lightweight, low atomic number gas to push equally against both the neutron star surface and the contact surface, that is, the piston, of the ejection shock. Such a lightweight gas is the photon-pair-dominated gas that forms at high temperature, provided it is nucleon deficient. A measure of this deficiency is that there must be at least 100 photons with their associated electron pairs per nucleon such that the energy density per nucleon is greater than the nucleon binding energy to the neutron star or hence, entropies of the order of 100 to 1000 at $T \simeq 1$ MeV ($\rho_{nucleon} \leq 10^6$ g cm^{-3}). This is just the entropy range formed in the hot bubble in Wilson's and Mayle's recent calculations.

Older calculations of Wilson,[5] and in Bethe and Wilson,[6] showed a weaker hot bubble, $s \leq 100$, whose existence was dependent sensitively upon input parameters. Instead, the present calculation, which utilizes a more exact treatment of neutrino–neutrino annihilation, shows a much higher entropy, larger hot bubble. The hot bubble is aided by the bounce shock, which temporarily reduces the flow of matter onto the neutron star. At the same time, the neutrino emission that carries the heat of the binding energy of the neutron star is emitted in neutrinos in all flavors: electrons, muons, and taus and their antiparticles. The density of nucleons

($\rho_{nucleon} \leq 10^9$ g cm^{-3}) following the bounce shock is not in itself enough to absorb these neutrinos and cause it to heat into a hot bubble. Instead, the neutrino number density itself from the heat flux is far greater than ($\times 100$) the nucleon component, and as was first shown by Goodman et al.,[2] these neutrinos and particularly the very hottest neutrinos, the mus and taus ($T \simeq 10$ MeV) that carry the most energy, annihilate on each other into electrons that create the hot bubble ($\nu_{\mu,\tau} + \bar{\nu}_{\mu,\tau} \to e^+ + e^-$). The energetic electron pairs immediately thermalize, resulting in a photon plus pairs plus nucleon gas. As Wilson and Mayle[1] have discussed, the extreme energy dependence of this neutrino–neutrino annihilation process requires that the neutrinos preferentially annihilate in nearly head-on collisions, or at least with extreme angular dependency of the order of the eighth power of the angle. Hence, only when the neutron star surface has become extremely sharp, that is a very small-scale height at the mu–tau neutrino photosphere, will the neutrino–neutrino annihilation start heating the external low-density matter to form the bubble. This happens some few tenths of a second following the bounce, and thus forms a hot cavity that expands rapidly, driving the ejection shock. Once the hot bubble starts to form, the pairs, associated with the high-temperature gas, become the dominant particle for neutrino interaction, and so once started, the hot bubble is further heated by the entire neutrino flux, only a small fraction of which has annihilated to form the initial hot bubble. This further heating produces, at the end of 3.6 s, a hot bubble whose entropy is as high as 10^4 in places and whose temperature adjacent to the contact surface, that is, the piston driving the shock is now low enough, a few hundred keV, that its subsequent cooling by neutrino emission is negligible.

MIXING BEHIND THE SHOCK

The consequence of the hot bubble is that it will mix at the contact surface of the expanding shocked matter. This shocked matter has considerably lower entropy than the hot bubble. To investigate the degree of this mixing there are several surprising simplifications to the problem. These are (1) the expected mixing is close to the thickness of matter behind the shock, and (2) the shock entropy increases due to the decreasing density of the envelope and the two entropies: that of the expanding hot bubble and that of the material behind the shock will become equal somewhere near the boundary between helium and hydrogen or when $\rho \simeq 1$ g cm^{-3}. Thus, the outer mass fraction of ejected hydrogen will be unmixed, and inside everything will be mixed.

TURBULENT MIXING

Read[7] and Youngs[8] developed measurements and theories that show that the Rayleigh Taylor unstable mixing leads to a turbulent boundary layer separating the two fluids. This is the nonlinear limit of growth from a thermal noise spectrum of an initial interface with no other initial perturbations and infinite Reynolds numbers. They found that the thickness grew as a function of acceleration and time as

$$\Delta x = 0.07at^2, \tag{1}$$

where a is the acceleration. Since the displacement of the interface increases as

$$x = \frac{1}{2at^2},\tag{2}$$

the thickness of the mixed layer becomes $\frac{1}{7}$ of the displacement. As an aside, the mechanism of growth is that the spikes and bubbles of the nonlinear limit of the dominant growing wavelength excite eddies of equal scale, some of which coalesce by inverse turbulent cascade and thereby excite the next larger wavelength perturbation, etc. In the case of the stellar shock, the slowing down of the contact surface between the hot bubble material at higher entropy and the shocked matter at lower entropy is what drives the instability. This slowing down is due to the increasing mass of the matter external to the shock. In general, the effect of gravity is less than the deceleration due to increasing mass behind the shock, but contributes at the early phase of the expansion.

ACCELERATION OF CONTACT SURFACE

We assume that the hot bubble has been formed and expands adiabatically. Both the hot bubble and the material behind the shock are radiation dominated, and so $\gamma = \frac{4}{3}$. Hence, in a homologous expansion no pressure gradient is formed across the contact surface due to different adiabats.

The acceleration of the contact surface, then, is due to a decreasing shock velocity and gravity. The contact surface is always close to the shock, $\Delta R \simeq R/(3\eta)$ because the compression ratio

$$\eta = \frac{\gamma + 1}{\gamma - 1} = 7$$

is large and the entropy gradient is weak.

The density distribution is close to polytropic external to the core and so to good approximation

$$\rho = \rho_0 \left(\frac{R}{R_0}\right)^{-3},\tag{3}$$

where $\rho_0 = 4 \times 10^4 \, \text{g cm}^{-3}$ at $R_0 = 2.5 \times 10^9$ cm. This gives a logarithmically increasing mass with radius or $M = 26.5 M_\odot$ at a surface $R = 2 \times 10^{12}$ cm. The density falls slightly faster than r^{-3}, resulting in half this mass, but Eq. (3) is a sufficient approximation for the inner regions. With this distribution there also exists matter inside R_0 of the hot bubble that has collapsed to the neutron star. The Wilson–Mayle calculations put this mass at $1.63 M_\odot$. If we take $10 M_\odot$ ejected with 1.5×10^{51} ergs—typical of the models (Woosley,[9] Arnett,[10] Nomoto et al.[11])—then the surface velocity of the equivalent uniform density sphere is

$$\frac{u_{\text{surf}}^2}{2} \frac{5}{3} M_{ej} = 1.5 \times 10^{51} \text{ ergs},\tag{4}$$

or $u_{surf} = 3 \times 10^8$ cm s^{-1}. The fluid velocity behind the ejection shock that gives rise to this free surface velocity is closely one-half of the surface velocity, contrary to simple energy conservation (Colgate and White[12]). Hence, the fluid velocity behind the shock near the surface is $\simeq 1.5 \times 10^8$ cm s^{-1}, neglecting the speed up of the shock in the final density gradient. The corresponding velocity at the radius of $R_0 = 2.5 \times 10^9$ cm will be larger by the square root of the internal mass ratio or by $(\ln R_s/R_0)^{1/2} = 2.6$ in order to conserve energy. Thus, $u_0 = 4 \times 10^8$ cm s^{-1}. The corresponding internal energy of the hot bubble is then $3P_0 \times$ Vol, where $P_0 = \rho_0 u_0^2 \times 7/6 = 7.4 \times 10^{21}$ dyne cm^{-2}, or $W_{bubble} = 1.4 \times 10^{5'}$ ergs. The sum of internal and kinetic energy behind the shock at $u_0 = 4 \times 10^8$ cm s^{-1} and $(M_0 = 1.6 M_\odot)$ is 0.5×10^{51} ergs or a total of 1.8×10^{51} ergs. Roughly 0.8×10^{51} ergs is retained as binding energy, giving the self-consistent value of final kinetic energy of 1×10^{51} ergs.

ENTROPY AND MIXING

Using this model we can calculate the entropy. If we use the variable $s = P/\rho^{4/3}$, then for a bubble entropy of 10^2, the minimum value necessary to retain a pressure against the neutron star in units of k, the Boltzmann constant, $s_b = 1.4 \times 10^{17}$. The entropy behind the shock in the same units is

$$s_0 = \frac{\rho_0 u_0^2 \eta/(\eta - 1)}{(\rho_0 \eta)^{4/3}} = 8.7 \times 10^{-2} \frac{u_0^2}{\rho_0^{1/3}} = 4 \times 10^{14}. \tag{5}$$

Thus, the shock entropy and the bubble entropy become equal when

$$s_b = s_{shock} = s_0 \left(\frac{\rho_s}{\rho_0}\right)^{1/3} \ln \frac{R_s}{R_0}. \tag{6}$$

Since $(\rho_s/\rho_0)^{1/3} = R_s/R_0$, we obtain a radius $R_s = 80 R_0 = 2 \times 10^{11}$, or where the stellar envelope density is $\rho_s = 0.05$ g cm^{-3}.

The integral of the acceleration determines the degree of mixing up to this point in the envelope. If the velocity had decreased to zero in this distance and neglecting gravity, the mixed layer would be 1/7 the radius. The velocity decrease is only-half, however, and so the acceleration is reduced in half, so that roughly the mixed zone is only 1/2 or $R/14$. On the other hand, a uniform density sphere of matter compressed by η will appear as a layer $\Delta R/R = (1/3\eta)$ thick or $R/21$. Hence, a strong shock will be compressed to a thin layer in a uniform medium, and be somewhat thicker with our density distribution $\rho \propto R^{-3}$. Thus, the mixing will penetrate close to the shock. A decreased mixing can also be expected due to the decreasing entropy of the bubble itself due to mixing with the lower entropy shocked matter. Thus, more detailed calculations are warranted, but it appears that one can expect the hot bubble to mix out to roughly the helium–hydrogen zone in the star, $\rho \simeq 1$ g cm^{-3}, and possibly somewhat further. This is just what is required to give the gamma-ray transparency from ^{56}Ni $\rightarrow$ ^{56}Co $\rightarrow$ ^{56}Fe decay.

It is noteworthy that the astrophysical attempt to understand the explosion of supernovas using the calculational tools developed for nuclear weapons is now a tradition of 30 years at Lawrence Livermore Laboratory (LLL) (Colgate and

White[12]). The existence of the Z particle, which permits neutrino–neutrino annihilation, was not known at that time. Demonstrating the solution to the problem at LLL is a tribute to this long dedication.

REFERENCES

1. WILSON, J. R. & R. MAYLE. 1984. *In* Proceedings of the NATO Conference in the Nuclear Equation of State. Springer-Verlag. Berlin/New York.
2. GOODMAN, S. A., A. DAR & S. NUSSINOV. 1987. Astrophys. J. **314:** L7.
3. COLGATE, S. A. 1971. Astrophys. J. **163:** 221.
4. BONDI, H. 1952. Mon. Not. R. Astron. Soc.
5. WILSON, J. R. 1985. *In* Numerical Astrophysics, J. Centrella, J. LeBlanc, and R. Bowers, Eds.: 422. Jones and Bartlett. Boston.
6. BETHE, H. & J. WILSON. 1985. Astrophys. J. **295:** 14.
7. READ, K. I. 1984. Physica 12D: 45–58. North-Holland. Amsterdam, The Netherlands.
8. YOUNGS, D. L. 1984. Physica 12D: 32–44. North-Holland. Amsterdam, The Netherlands.
9. WOOSLEY, S. E. 1988. Astrophys. J. **330:** 218.
10. ARNETT, W.D. 1988. *In* Supernova 1987A in the Large Magellanic Cloud, M. Kafatos and A. Michalitsianos, Eds.: Cambridge University Press. Cambridge, England.
11. NOMOTO, K., T. SHIGEYAMA & M. HASHIMOTO. 1987. *In* SN 1987A, I. J. Danziger, Ed.: 325. ESO, Garching.
12. COLGATE, S.A. & R. H. WHITE. 1966. Astrophys. J. **143:** 626.

Thermonuclear Ignition and Runaway in Type Ia Supernovae[a]

J. CRAIG WHEELER

Department of Astronomy
University of Texas at Austin
Austin, Texas 78712

INTRODUCTION

One of the most common supernova events, classified as Type Ia (SN Ia, Wheeler and Harkness[1] and references therein), is thought to arise not in a gravitational collapse, but in a thermonuclear explosion. The study of these events brings in a large range of problems in nonlinear dynamics from convective burning to the final explosion.

SN Ia are thought to arise in the explosion of degenerate carbon/oxygen (C/O) white dwarfs. The actual evolutionary path leading to this outcome is not well understood. Because they occur in elliptical as well as other kinds of galaxies, they are associated with older populations. The qualitative picture is that mass transfer in a binary system brings a C/O white dwarf to very near the Chandrasekhar limit, at which point it undergoes runaway carbon burning. Wheeler[2] gives a summary of the problems involved in constructing a quantitative model of the required evolution.

In this paper the story is picked up at the time of carbon ignition. The second section discusses the problem of the convective Urca process, which will delay the immediate runaway. The question of detonation versus deflagration in the subsequent dynamics is discussed in third section. The last section illustrates how the light curves and spectra can be used to constrain models of the dynamics.

IGNITION: THE CONVECTIVE URCA PROCESS

As will be shown in the final section, the models that best match the observations of SN Ia involve the explosion of C/O white dwarfs. The initial configuration is assumed to be a degenerate star with a central density of $\rho_c \sim 2$–$3 \times 10^9 \, \mathrm{g \, cm^{-3}}$ and a central temperature of $T_c \sim 3 \times 10^8$ K. These are the conditions when the rate of thermonuclear burning in carbon exceeds the local neutrino loss rate. The corresponding mass is very near to the Chandrasekhar limit. After the ignition of carbon the star begins a phase of convective carbon burning during which the convection can control the release of energy (Arnett[3]). Because some form of accretion is presumed to continue to feed the growth of the core, the central density and temperature will increase. If nothing else intervenes, the nuclear rates will increase until the convection cannot control the burning, and a dynamical runaway will ensue. In practice,

[a]This research was supported in part by NSF Grant 8717166 and in part by a grant from the R. A. Welch Foundation.

8

another complex process comes in that makes the subsequent quasistatic evolution difficult to analyze.

This process is called the *convective Urca process*. It was first suggested by Paczyński[4] as a generalization of the basic Urca process of Gamow and Schoenberg.[5] Paczyński argued that the convective core will sweep β unstable nuclei around a cycle from high to low densities. At the high density there will be electron capture with the generation and loss of a neutrino, and at low density a β decay, with the loss of an antineutrino and the restoration of the original composition. Paczyński showed that the rate of neutrino losses is sensitive to the degree to which the convective core extends beyond the threshold for the β processes. The effective temperature sensitivity of the neutrino losses is of order T^{170}. This is much more sensitive than the nuclear burning that goes approximately as T^{20}, and so the Urca neutrino process can, in principle, control the burning.

Bruenn[6] pointed out that the electron capture occurs below the Fermi sea, so that subsequent cascades can release heat. Beta decay also produces heat, and Bruenn argued that the convective Urca process could lead to net heating, and hence no effective change in the carbon runaway process. Couch and Arnett[7] argued that the mechanical energy of convection must be included in the prescription. They assumed cooling by the convective Urca process and concluded that runaway could be postponed until $\rho_c \geq 6 \times 10^9 \, g \, cm^{-3}$.

The most thorough numerical study of the convective Urca process has been done by Iben[8] (and references therein). Iben found that his model went through a series of thermal cycles of growing amplitude driven by the $^{23}Na-^{23}Ne$ pair. When the density got high enough to trigger the $^{21}F-^{21}Ne$ pair, a new Urca shell was formed in the interior, isolating the outer shell. The outer shell could not then control the burning within it and the burning ran away, despite the inner Urca shell. Iben found the runaway to occur at $\rho_c \geq 4 \times 10^9 \, g \, cm^{-3}$. This density is not high enough to lead to a dynamical collapse, but it is high enough to cause problems with the neutronization in the resulting ejecta. Due to the complexity of Iben's computations, they were not fully assimilated, or perhaps understood, by the community. Most subsequent calculations of degenerate carbon runaway have ignored the convective Urca process and assumed runaway at 2–3 $\times 10^9 \, g \, cm^{-3}$.

The physics of the convective Urca process has been reexamined by Barkat and Wheeler.[9] They argue that in steady state there must be gradients in the composition profiles of the Urca nuclei, whereas constant composition had been assumed in some work. They have shown that when careful account is taken of the chemical potential currents of the Urca-active species that attend the composition gradients, the heating terms discussed by Bruenn[6] are exactly canceled in steady state and the result is cooling, just as originally described by Paczyński.

Barkat and Wheeler[9] further argue that the state of balance achieved by the cooling Urca process will respond in an unsymmetric way to entropy perturbations, and that the net result is that the equilibrium must be unstable. They argue that if the entropy of the convective core is perturbed upward, the convective core will expand outward. The Urca cooling will overwhelm the nuclear heating and thus stabilize such a perturbation. The response to a negative perturbation to the entropy is more complex. If the convective core were not to shrink in response to a negative entropy pertubation, the Urca cooling would remain strong, but the nuclear rates would decrease. This would lead to rapid, unstable cooling. In practice, one expects the convective core to shrink in response to a negative entropy perturbation. This will decrease the extension of the convective core beyond the Urca shell, and hence both the Urca losses and the nuclear burning. The quantitative result will depend on physical details concerning the rate of outward motion in mass of the Urca shell due to the increase in the core mass and density and of the shrinking of the convective core.

Barkat and Wheeler[9] argue that the core will always be subjected to a negative entropy pertubation because of the "ordinary" neutrino losses that operate immediately beyond the convective Urca core. When the cooling time scale is less than the core growth time scale, then thermal cycles are the expected result. As the core undergoes unstable cooling the convective core must eventually shrink until that coupled with the rate of advance of the Urca shell decreases the Urca cooling. At that time nuclear reactions will result in heating that will drive the system back to the condition of unstable thermal balance and the cycle will repeat. Such cycles may be the origin of the thermal oscillations reported by Iben.[8] The subsequent behavior including the development of a new central Urca shell probably merits a careful, independent numerical study.

PROPAGATION: DETONATION OR DEFLAGRATION

It is not completely clear that the Urca process will not postpone the full thermonuclear runaway of carbon to very high densities. For $\rho_c \geq 2 \times 10^{10}$ g cm^{-3} weak interactions in the burned matter can occur on a dynamical time scale and induce collapse. Nevertheless, the most careful studies to date (e.g., Iben[8]) suggest that dynamical runaway will occur at more moderate central densities and that the result will thus be complete disruption of the star.

The nature of the dynamical event has been debated for two decades. Arnett[3] argued that the likely result of thermonuclear runaway was to form a supersonic detonation. In this process the burning drives a shock that triggers subsequent burning. The steady-state self-driven condition is called a Chapman–Jouguet detonation. In contrast, Nomoto, Sugimoto, and Neo[10] and Chechetkin et al.[11] argued that due to the weak overpressures generated in the burned matter, spherical divergence, and other effects, a subsonic deflagration driven by conductive or convective heat transport was more likely. It is now recognized that these numerical studies were compromised by insufficiently refined zoning and probably have little direct bearing on the problem. Mazurek, Meier, and Wheeler[12] attempted to analyze the problem semianalytically using techniques appropriate to shock tubes. In particular, they examined the effects of accumulated shocks as subsequent portions of the star burned. They concluded that due to the weak overpressure (~ 10–20 percent) in the initially highly degenerate matter, the accumulated effects of a series of isochoric burning events was insufficient to develop the postshock pressure of a Chapman–Jouguet detonation. Mazurek et al. thus deduced that a detonation could not develop and that some form of deflagration would result. See Mazurek and Wheeler[13] for a summary of the dynamics of stellar detonations and deflagrations.

While it was not clear that this was the final answer, the issue was then strongly influenced by observations. SN 1981b in NGC4536 was the brightest, best studied SN Ia in nearly a decade. The first successful spectral synthesis calculations (Branch *et al.*[14,15]) showed that the atmosphere near maximum light contained large abundances of intermediate mass elements such as O, Mg, Si, S, and Ca. Since detonations are supersonic, the outer portions of the star remain unperturbed until the burning front arrives. The structure remains compact, and this promotes the propagation of the burning essentially to the surface of the C/O core. The result is to consume virtually the entire core to nuclear statistical equilibrium. The ejecta will be nearly all iron-peak matter with only negligible amounts of intermediate mass elements as velocities far in excess of those observed. In contrast, a deflagration evolves from a transient isochoric burning phase that sends a precursor shock out through the core, preaccelerating it to finite velocity. This is necessary to the kinematics because the deflagration involves rarefaction in its wake, and some expansion of the central regions to lower densities is necessary so that the deflagration can occur. As the subsonic deflagration propagates, other shocks are continually emitted that add to the preacceleration of the outer, unburned matter. The natural result is that some of the outer material has expanded to densities $\leq 10^8$ g cm^{-3} before the burning front arrives. Below this density the nuclear burning time scale becomes longer than the dynamical expansion time scale, and the burning is quenched. A region of partially burned matter generated as the burning fades is thus a generic feature of such models, independent of the ill-understood details of precisely how, or at what speed, the deflagration front propagates (Nomoto, Thielemann, and Yokoi[16]). Although the details have changed and the atmosphere calculations have become more sophisticated, the results remain much the same. As will be shown in the next section, the deflagration models give basically good agreement with the observed spectra (Branch *et al.*[17]; Harkness[18-20]). The deflagration models in the literature match the observations near maximum light; those involving detonations do not. The same is true for the later, nebular phases (Woosley, Axelrod, and Weaver[21]).

Nevertheless, Mazurek, Meier, and Wheeler[12] overlooked a major component of the problem. They neglected to consider the motion of the matter subsequent to the stage of isochoric burning. The expansion of the burned matter serves to accelerate the strength of the shock, and this strongly promotes the development of a detonation. This was first realized by Blinnikov and Khokhlov.[22,23] They examined the problem of degenerate carbon burning dynamic runaway on mass scales $\leq 10^{-5} M_\odot$ at which one can resolve the spatial propagation of the burning front. For a finite temperature gradient, a finite region burns to completion on a dynamical time scale. Within this region of spontaneous burning all mass elements burn to completion even though there is no sonic communication between them. The effective phase velocity of the burning front is $v_{phase} > c_s$, the local sound speed. Blinnikov and Khokhlov concluded that this region of spontaneous burning and the subsequent kinematics would tend to promote detonations. Related analyses, begun independently, were given by Woosley and Weaver[24] and Wheeler *et al.*[25] Woosley[26] concludes that a deflagration is still not ruled out, but Barkat[27] points out that Blinnikov and Khokhlov neglected the finite burning that will occur in the region just beyond the region of spontaneous burning. He notes that at the boundary of the spontaneously burned region the pressure and postfront velocity are precisely those of a Chapman–Jouguet detonation, although a shock wave has not yet formed. Considering the finite burning beyond this boundary, Barkat concludes that a detonation is even more likely than thought by Blinnikov and Khokhlov.

It is difficult to establish the central temperature at runaway below which a detonation will not form because at sufficiently low temperatures the length scale to evolve a full-fledged detonation becomes longer than the total mesh size of a sufficiently finely zoned calculation. Barkat[27] concludes that a central temperature as low as 6×10^8 K for the initial runaway will still lead to a detonation, and this is less than the central temperature at dynamical runaway for many numerical calculations.

Trial calculations with ignition initiated off-center in a radial grid still develop detonations, so it is not clear that multidimensional effects will radically alter these conclusions. In a realistic situation the initial temperature structure will be irregular due to the preexisting convective core. Whether steep temperature gradients at the boundaries of convective elements will damp the tendency to detonate or whether small nearly isothermal pockets will promote detonation is not clear.

The net result is that the physics of the runaway tends to point strongly in the direction of detonations. If this situation persists, there is a problem of reconciling the physics with the observations of partially burned matter in SN Ia explosions. One possibility that requires exploring is a model in which for some reason the core expands before the full-fledged detonation is generated. This could lead to the development of more extensive outer regions that were too low in density to support a detonation by the time it arrived. Such a situation might occur if an isolated hot spot ran away near the center of the star, providing an initial, but limited, input of nuclear energy. As other spots burned and their fronts overlapped, a detonation could develop that would nevertheless be quenched in the expanding outer portions of the core. Such a quenched detonation might prove very difficult to distinguish from a deflagration.

DIAGNOSTICS OF THE EXPLOSION

The question of whether a damped detonation or a deflagration is the best model for an SN Ia is apt to be decided by observations rather than by argument from first principles of physics. There are a number of ways to constrain the explosion. The primary tools are the light curves and spectra. In special cases one can also probe the density profile directly.

There are relatively few detailed numerical models of deflagrations in the literature. Among these, the most studied are those presented by Nomoto, Thielemann, and Yokoi.[16] Others have been given by Woosley and Weaver.[24] These models are all qualitatively the same, but there are some interesting and important differences. The models yield an inner region of iron-peak material. Since the deflagrations expand relatively slowly, they have a tendency to undergo more weak interactions in this matter during the expansion, and the result is a relative excess of neutron-rich species compared to iron. Woosley[26] argues that this is also a problem with detonation models. How serious it is ultimately depends on the rate of explosion of such events. If they are as infrequent as once per 300 years in the Galaxy

(Wheeler[28]) and Type II supernovae explode about one per 50 years and dominate the nucleosynthesis output, then the problem of overproduction of neutron-rich species is muted.

The deflagration models also give the important region of partial nuclear burning as the front dies out. The deflagrations in the models of Nomoto et al.[16] are modeled in such a way that the ratio of front speed to the sound velocity increases with time. They find a relatively extended region of partial burning. The models of Woosley and Weaver[24] have deflagrations that propagate at approximately a constant fraction of the sound speed. In these models, the partially burned zone is rather abbreviated in mass and velocity. In general, the deflagration models then also have an outer region of unburned carbon and oxygen. Nomoto et al. presented a model (C6) with a relatively slow deflagration speed. This model produced 0.48 $M_\odot$ of ^{56}Ni, which decays to power the light curve and left 0.4 $M_\odot$ of unburned C/O. A somewhat faster model (W7) produced 0.58 $M_\odot$ of ^{56}Ni. The burning front propagated nearly to the surface of the star, producing the intermediate mass elements at rather high velocities and over a rather large velocity range. Only 0.08 $M_\odot$ of C/O remained unburned. For a slightly faster deflagration speed (Model W8) the deflagration evolved into a detonation and produced virtually no intermediate mass elements. Model G7 of Woosley and Weaver produced 0.51 $M_\odot$ of ^{56}Ni and 0.4 $M_\odot$ of C/O. Their Model F7 with a faster deflagration produced 0.89 $M_\odot$ of ^{56}Ni and no unburned C/O. The special feature of this model was that after the deflagration died, the remnant shock accelerated to a detonation in the outer layers, turning them entirely to iron-peak elements.

Light Curves

The light curves of these models and variations on them have been explored with a uniform set of assumptions for the γ-ray deposition function and optical opacities by Boisseau.[29] Models W7 of Nomoto et al.[16] and G7 of Woosley and Weaver fit a composite SN Ia bolometric light curve[30] within the scatter. Model C6 tends to drop too slowly in the first 60 days because the ^{56}Ni lies deep in the model so that the γ-rays are efficiently trapped for a longer time. On the contrary, Model F7 tends to drop a little too rapidly because the ^{56}Ni resides farther out in the matter so the γ-rays can escape more easily.

Trial calculations in which the ^{56}Ni is moved around (not self-consistently) in the model show that the light curve is somewhat sensitive to the distribution as argued by Graham.[30] Our models show, however, that the difference between the original Model W7, in which the ^{56}Ni falls in an off-center shell, and a version in which all the ^{56}Ni is arbitrarily moved to a central core, is within the scatter of the observational curve. Graham made a number of simplifying assumptions that gave a larger difference in these two configurations.

We conclude that although light curves can provide a basic discriminant between models that differ in major ways, they are insufficient to differentiate more subtle differences. Light curves would nevertheless be a better tool if there were a number of well-established bolometric light curves for SN Ia. That is not the case today.

Spectra

LTE radiative transfer calculations have been quite successful in reproducing semiquantitatively the near-maximum-light optical and ultraviolet spectra, including the steep UV deficit of SN Ia (Harkness,[18-20] Harkness and Wheeler,[31] Wheeler and Harkness[1]). Harkness[20] shows, however, that of the deflagration models discussed in the previous section, only Model W7 provides an adequate agreement with observations. Model C6 of Nomoto et al.[29] is too slow and the spectral features too sharp, although they are qualitatively correct. Model G7 of Woosley and Weaver[24] has the intermediate mass elements buried too deeply in mass at too narrow a velocity range. The resulting maximum light spectrum is scarcely qualitatively recognizable. Model F7 with the detonated outer layers has basically no resemblance to observed spectra at all.

Harkness[20] has also shown that with improvements in the opacity, for example, treating weak lines as a quasicontinuum in the local rest frame, good agreement is obtained between the basic W7 model and observations with no need to invoke an *ad hoc* mixing and homogenization of the outer layers, as has been done in some previous calculations.[17,19]

Thus, generic deflagration models that more or less satisfy light-curve constraints do not necessarily match the spectra at all well. Model W7 with its partial burning that reaches essentially all the way to the surface is special in its ability to match the observations. This emphasizes the question of whether a damped detonation with the same property of partial burning nearly to the surface would not do a better job than most deflagration models do.

Other Constraints

The aspect of the Nomoto et al.[29] W7 model that gives agreement with spectral observations is that the deflagration accelerates so that the partial burning does not truncate precipitately, but extends over a rather broad mass and velocity range almost entirely to the surface. An interesting and presumably related aspect of this calculation is that the resulting density profile shows a peak in the very center ($M/M_{TOT} \lesssim 0.1$) and little evidence for a piled-up "shell" of matter in the outer, unburned regions. Nomoto's Model C6 also shows the central peak, but has a marked outer shell. By contrast, the deflagration models of Woosley and Weaver[24] yield a density profile in the inner half mass fraction that is homologous to the original white-dwarf static-density profile. The central peak in the Nomoto model is presumably related to the slow speed of the deflagration in the center. It may be that subsequent electron capture reduces the pressure and allows the density to increase in the center, but this is not established.

Hamilton and Fesen[32] have provided a novel way of constraining the density profile. They have analyzed Fe absorption lines from matter in SN 1006, a putative SN Ia remnant, against a background subdwarf. They model the Fe II absorption profiles by computing the interaction of the expanding ejecta of a supernova model with the ISM and the resultant production of ionizing X rays. They find that Model G7 of Woosley and Weaver[24] gives a rather good representation of the deduced Fe II density/velocity profile, whereas Model W7 of Nomoto et al.[29] does not. The problem is that without the outer shell, W7 reacts with the ISM rather weakly and produces a smaller ionizing flux. That smaller ionizing flux has an even more difficult time in ionizing Fe II into higher, unobserved ionization states, because the dense peak of matter in the center shields the Fe II from the ionizing flux. Thus, the very aspect of Model W7 that is required to agree with spectral observations, the partial burning extending to the surface, and the concomitant lack of the outer shell, are what cause it to fail the constraints of Hamilton and Fesen. The central density peak, to which the maximum light spectra are insensitive, is also an aspect of the model of deflagration propagation that is not favored by Hamilton and Fesen. They do get reasonable agreement with Woosley and Weaver's Model G7, which has an outer shell to generate copious X rays and no inner peak to shield the Fe II from ionization, but this model produces a very poor optical spectrum.

CONCLUSIONS

The convective Urca process is a complication to the problem of degenerate carbon ignition that must be faced. Either the density at which carbon first ignites must be changed rather severely (to $\rho_c \leq 10^9 \, \mathrm{g\,cm^{-3}}$) or the effects of convective Urca must be incorporated in detailed models. The results of Iben[8] with runaway in an off-center, unregulated Urca shell may be correct. Alternatively, dynamical runaways may be postponed even further. If Iben is correct, then models for dynamic runaway and the question of detonation versus deflagration should be posed in this ambiance, runaway in an off-center convective zone with the complex temperature distribution that configuration implies.

According to the few detailed models in the literature, one deflagration model provides (perhaps surprisingly!) a reasonably good agreement with spectral observations, but most do not. The model that agrees with the spectra does not agree with independent constraints on the density profile. This situation suggests that it is premature to conclude that deflagrations provide the solution to the SN Ia problem. Alternative solutions involving damped detonations should be explored.

REFERENCES

1. WHEELER, J. C. & R. P. HARKNESS. 1990. Rep. Prog. Phys. In press.
2. WHEELER, J. C. 1990. *In* Frontiers of Stellar Evolution, D. L. Lambert, Ed. Astronomical Society of the Pacific, San Francisco. In press.
3. ARNETT, W. D. 1969. Appl. Space Sci. **5**: 280.
4. PACZYŃSKI, B.E. 1973. Astrophys. J., Lett. **15**: 147.
5. GAMOW, G. & M. SCHOENBERG. 1941. Phys. Rev. **59**: 539.
6. BRUENN, S. W. 1973. Astrophys. J., Lett. **184**: L125.
7. COUCH, R. G. & W. D. ARNETT. 1975. Astrophys. J. **196**: 791.
8. IBEN, I., JR. 1982. Astrophys. J. **253**: 248.
9. BARKAT, Z. & J. C. WHEELER. 1990. Astrophys. J. **355**: 602.
10. NOMOTO, K., D. SUGIMOTO & S. NEO. 1976. Astrophys. Space **39**: L37.
11. CHECHETKIN, V. M., V. S. IMSHENNIK, L. N. IVANOVA & D. K. NADYOZHIN. *In* Supernovae, D. N. Schramm, Ed: 159. Reidel. Dordrecht, The Netherlands.
12. MAZUREK, T. J., D. L. MEIER & J. C. WHEELER. 1977. Astrophys. J. **215**: 518.
13. MAZUREK, T. J. & J. C. WHEELER. 1980. Fundam. Cosmic Phys. **5**: 193.

14. BRANCH, D., R. BUTA, S. W. FALK, M. L. McCALL, P. G. SUTHERLAND, A. UOMOTO, J. C. WHEELER & B. J. WILLS. 1982. Astrophys. J., Lett. **252:** L61.
15. BRANCH, D., C. H. LACY, M. L. McCALL, P. G. SUTHERLAND, A. UOMOTO, J. C. WHEELER & B. J. WILLS. 1983. Astrophys. J. **270:** 123.
16. NOMOTO, K., F.-K. THIELEMANN & K. YOKOI. 1984. Astrophys. J. **286:** 644.
17. BRANCH, D., J. B. DOGGETT, K. NOMOTO & F.-K. THIELEMANN. 1985. Astrophys. J. **294:** 619.
18. HARKNESS, R. P. 1986. *In* Radiation Hydrodynamics in Stars and Compact Objects, D. Mihalas and K.-H. A. Winkler, Eds.: 166. Springer-Verlag. Berlin/New York.
19. ———. 1987. *In* Relativistic Astrophysics, M. P. Ulmer, Ed.: 413. World Scientific. Singapore.
20. ———. 1990. *In* Supernovae. S. E. Woosley, Ed. Springer-Verlag. New York/Berlin. In press.
21. WOOSLEY, S. E., T. S. AXELROD & T. A. WEAVER. 1984. *In* Stellar Nucleosynthesis, C. Chiosi & A. Renzini, Eds.: 263. Reidel. Dordrecht, The Netherlands.
22. BLINNIKOV, S. I. & A. M. KHOKHLOV. 1986. Sov. Astron. Lett. **12:** 131.
23. ———. 1986. Sov. Astron. **13:** 364.
24. WOOSLEY, S. E. & T. A. WEAVER. 1986. *In* Radiation Hydrodynamics in Stars and Compact Objects, D. Mihalas and K.-H. A. Winkler, Eds.: 91. Reidel. Dordrecht, The Netherlands.
25. WHEELER, J. C., R. P. HARKNESS, Z. BARKAT & D. SWARTZ. 1986. Publ. Astron. Soc. Pac. **98:** 1018.
26. WOOSLEY, S. E. 1990. *In* Supernovae. A. Petschek, Ed. Springer-Verlag. New York/Berlin. In press.
27. BARKAT, Z., E. LIVNE, J. C. WHEELER & D. SWARTZ. 1990. *In* Supernovae. T. Piran, S. Weinberg, and J. C. Wheeler, Eds.: 33. World Scientific. Singapore.
28. WHEELER, J. C. 1990. *In* Supernovae. T. Piran, S. Weinberg, and J. C. Wheeler, Eds.: 1. World Scientific. Singapore.
29. BOISSEAU, J. 1989. Unpublished.
30. GRAHAM, J. R. 1987. Astrophys. J. **315:** 588.
31. HARKNESS, R. P. & J. C. WHEELER. 1990. *In* Supernovae. A. Petschek, Ed.: 1. Springer-Verlag. New York/Berlin.
32. HAMILTON, A. J. S. & R. A. FESEN. 1988. Astrophys. J. **327:** 178.

Chaotic Pulsations in Stellar Models[a]

J. ROBERT BUCHLER

Physics Department
University of Florida
Gainesville, Florida 32611

INTRODUCTION

It is now close to four centuries since the first recorded observations of the variability of *o* Ceti (Mira), and the study of variable stars has attracted a great many scientists ever since that discovery. Stellar pulsations are one of Nature's curiosities. For a physicist its numerous and wondrous ways of building such *natural heat engines*[1] are intriguing. Indeed, the intrinsic variable stars come in a variety of forms, ranging from the relatively regular classical Cepheids, to the semiregulars, to the irregular variables, not to mention the fascinating newcomers in the variable star club, namely the nonradial pulsators, such as the Sun and the white dwarfs.[2] I will discuss here only the behavior of the large-amplitude pulsators that undergo *radial* oscillations. At the present time a nonlinear study of nonradially pulsating stars is not feasible.

The physical mechanism of the pulsations is well understood:[3] it is the partial ionization regions of hydrogen and helium that provide the phase shift between the heat flux and the pulsation that is necessary to destabilize the static equilibrium, and nonlinear effects later saturate the linear instability.

Since the pioneering numerical hydrodynamic computations of Christy[4] in the early 1960s a great deal of effort has been directed toward the regular pulsators, viz., the classical Cepheids, the RR Lyrae, and a few others. The bulk nonlinear pulsational behavior of these objects is now well understood, although some very disturbing discrepancies remain.[5,6]

Theorists have devoted very little attention to understanding the physical origin of the *irregular* nonlinear behavior, although a good bit of data had been amassed by the observers, even prior to World War II. It is true that some people had suggested *dei ex machina* such as convective feedbacks,[7] unphysical modal spectra,[8] or had invoked mechanisms *extrinsic* to the star. The fact is that until a couple of years ago no compelling explanation had been offered.

Period doubling and chaos have been found to be a common occurrence in many fluid dynamics experiments,[9] and one would expect the same phenomenon to occur in astrophysical fluids. Indeed, there are at least some superficial indications of period doublings and chaos in the observational data. The stars classified as RV Tauri are characterized by a more or less irregular alternation of shallow and deep minima.[10] It is possible to assign an approximate period to these oscillations, but they are certainly not periodic. Observationally, the RV Tauri stars are contiguous to the longer period stars of the W Virginis group,[11] and they have very similar properties

[a]This research was supported by the National Science Foundation, the Pittsburgh Supercomputing Center, and the NER Data Center at the University of Florida.

aside from their different pulsational behavior. It is therefore not unnatural to suggest that they are the same type of star and that their pulsational behavior gradually changes from regular to irregular with increasing period, that is, with increasing luminosity and decreasing effective temperature.

Unfortunately, the available observational data do not readily lend themselves to an analysis with the modern methods that have been developed for chaos. At the present time we therefore have to rely on numerical hydrodynamic modeling. Even so, because of the high cost of this modeling, for the interpretation of the results we are forced to make use of the experience gained from laboratory experiments or simple numerical models. It should be obvious, however, that the complexity of the stellar pulsations exceeds by far that of the laboratory fluid experiments, on the one hand, and of the usually studied low-dimensional systems of equations, such as the Lorenz attractor,[12] on the other.

Irregular behavior is commonly classified as *chaotic* or *noisy,* depending on the number of degrees of freedom that are involved. Chaos is said to occur when the latter is small, and its interest lies in the fact that one can hope to find a *mechanistic,* as opposed to a *stochastic,* simple physical model for the observed behavior and, thence, gain a deeper understanding of the underlying dynamic.

It is interesting to note that RV-Tauri-like or irregular behavior in stellar models has been sporadically reported in the literature,[13-20] but that it has never received systematic attention, perhaps for want of a theoretical model within which to interpret such behavior.

Instead of performing a survey of the pulsational behavior of stellar models with a buckshot pattern in stellar parameter space, we have opted for a systematic study of *sequences* of stellar models. These sequences are constructed to be *one-parameter families* in which the luminosity L, the mass M, and the composition are fixed, and the effective temperature T_{eff} is the control parameter. The study of such sequences allows a systematic mapping of the bifurcation set, that is, of the different types of pulsations that are possible.

We mention in passing that the systematic study of sequences has already been found to be very fruitful in the study of Cepheids and RR Lyrae[21,22] models. In particular, in conjunction with the amplitude equation formalism[23-25] it has allowed a deeper understanding of the Hertzsprung bump progression (cf. Moskalik, Buchler, and Kovács in this volume), which is determined by a 2:1 resonance between the fundamental mode and the second overtone.[26-29]

In fact, when resonances are present, the shapes of the light and radial velocity curves can undergo rapid variations with L, M, and T_{eff}, and a survey with more or less randomly chosen stellar parameters may give a very confusing picture. Similarly, in a resonance region an attempt to model a specific star is fraught with risk, not only because its stellar parameters, in particular T_{eff}, are not known accurately, but also because, in our opinion, the hydrodynamic codes are not yet accurate enough for such a specific purpose.[30] On the other hand, when a survey is conducted with sequences of models we get an overall picture of the possible types of pulsation and of their appearance. Hopefully, when better modeling becomes possible this picture will remain essentially unaltered, but will probably be shifted to slightly different stellar parameters. Experiments with refined zoning and timesteps seem to corroborate this expectation.

PULSATIONAL BEHAVIOR OF THE MODEL SEQUENCES

The sequences to be discussed here are the W Virginis subgroup of the Population II Cepheids. These are stars that exhibit regular as well as RV Tau behavior. Our hydrodynamic studies have been performed with a slightly modified version of Stellingwerf's[31,32] implicit Lagrangean hydrocode[33] based on the Fraley[34] scheme. The equation of state includes the various ionization states of hydrogen and helium. Heat transport is treated in a diffusion approximation. In order to achieve smoothness along the sequences we use Stellingwerf's[32] fitted formula for the opacity. Most of the models described here are purely radiative. At the end we will describe some results obtained with time-dependent convection.

Our survey[35–37] which has consisted of a number of W Virginis sequences, has proceeded as follows.

1. A sequence of static-equilibrium models at fixed L, M, and composition is computed for a set of T_{eff}.

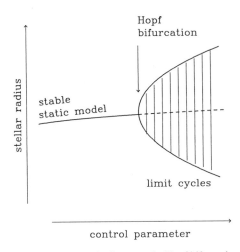

FIGURE 1. Schematic diagram of a Hopf bifurcation.

2. A linear "normal" mode analysis is performed assuming a temporal behavior of the form $\exp(i\sigma_k)$, which yields the spectrum $\{\sigma_k = i\omega_k + \kappa_k\}$, where the $\omega_k = 2\pi/P_k$ are the oscillation frequencies, and the κ_k are the linear growth or decay rates. The spectrum has a real subset, the thermal and secular modes, and a complex subset, the vibrational ones.[23] In the cases of interest here the real modes are all stable and play no important role. For the radial pulsators the ω_k are generally well separated (since the problem is of Sturm–Liouville type in the adiabatic limit). Simplifying a little bit, we sketch a typical scenario in FIGURE 1, where we show a model bifurcating from a stable static-equilibrium state to a limit cycle (κ changes from negative to positive). Such a bifurcation is generally termed a *Hopf bifurcation*. In the astrophysics literature the locus of this Hopf bifurcation in the H–R diagram is called the *linear*

blue edge. We note in passing that the situation can be a little more complicated, when an additional limit cycle attractor already exists at such a bifurcation;[21,22] for example, in classical Cepheids and RR Lyrae at the linear blue edge of the fundamental the first overtone limit cycle attractor is already present.

3. The nonlinear pulsational behavior of a vibrationally unstable static model is followed with the hydrocode until the asymptotic behavior is achieved. The latter can be (a) a regular attractor, either a limit cycle in which only one mode is excited (Fourier spectrum with a single fundamental frequency and constant amplitude), or a more complicated multiperiodic cycle with several incommensurate frequencies, or (b) an irregular or *chaotic* attractor.

- The singly periodic limit-cycles can be computed exactly with a relaxation code,[31,33] and their stability can be determined through a Floquet analysis. In the other cases time integrations are performed until the pulsation is estimated to have settled on its asymptotic state. In the multiperiodic case this can be checked with a computation of the Fourier phases and amplitudes, but in the chaotic case it is more difficult to be sure that the transient state has subsided.

It has been known for a long time that the pulsations of the classical variable stars are only weakly nonlinear. Indeed, the numerically computed nonlinear pulsations of the models always bear a strong resemblance to one of the low-frequency linear normal modes (the fundamental mode or one of the first two overtones), and the (nonlinear) period is very close to that of the corresponding linear normal mode (≤ 1 percent for RR Lyrae, several percent for the classical Cepheids; for the RV Tau models, on the other hand, period changes of 1 to 20 percent can occur). This nonlinearity is weak in spite of the large amplitudes that can occur, viz., $\Delta R/R \simeq 1$–7 percent for RR Lyrae and 5–15 percent for the classical Cepheids. It is important to keep in mind that the relative amplitude of oscillation has no direct bearing on the nonlinearity (as J. Perdang likes to point out, ocean waves are highly nonlinear phenomena, yet $\Delta R/R < 10^{-6}$). A more quantitative estimate of the nonlinearity could be obtained if the limit cycle were expanded in terms of the linear eigenvectors. At the Hopf bifurcation of the fundamental mode, for example, where the limit cycle has infinitesimal amplitude, all the projections except that on the fundamental eigenvector are then zero, and, as the amplitude of the limit cycle grows, its projections onto the other modes achieve nonzero values, although the projection onto the fundamental remains the largest. It is this closeness of the nonlinear pulsation to a linear mode that allows one therefore to talk about, say, the "fundamental" or "first overtone," limit cycles.

Recently, Buchler and Kovács[22,28] have given a more quantitative description of this nonlinearity based on the *amplitude equation formalism.*[23–25,38] This dimensional reduction method assumes that a few normal modes are only weakly unstable and that the vast majority of the modes of the system are strongly damped and, hence, that they can only exist as short-lived transients that are of no astrophysical interest. The dynamic of the system therefore shrinks to a very low-dimensional subspace of phase space, the so-called *slow manifold* (e.g., at a Hopf bifurcation this manifold is tangent to the two-dimensional space spanned by the the two complex eigenvectors of the bifurcating linear normal mode). The amplitudes of the slow modes are then sufficient to describe the full behavior of the model; they can be considered generalized coordinates in terms of which any other variables, such as velocities, temperatures, and pressures, can be expressed. The temporal behavior of these amplitudes is in turn governed by amplitude equations. The coefficients of these amplitude equations and those that appear in the expressions relating the amplitude to the other physical variables, can in principle be derived *a priori* from the static equilibrium model. At the present time this still represents a Herculean task and has only been attempted in a limited way.[26] Buchler and Kovács have instead assumed these coefficients to be parameters and have, with excellent results, fitted the computed hydrodynamic behavior with the appropriate amplitude equations both for RR Lyrae[33] and for bump Cepheids[28] (cf. also the paper of Moskalik *et al.* in this volume for a review). The interesting result for our purposes here is that it is sufficient to keep only the lowest order nonlinearities, namely, up to cubic ones in the amplitude equations.

As pointed out, the amplitude equation formalism is based on an assumption of weak dissipation (or nonadiabaticity). It is well known, however, that the outer regions of the stellar envelope are strongly nonadiabatic, and one may wonder why this approach works. It turns out that the important small parameter is not the *local* measure of nonadiabaticity, but rather its *global* measure. [We note in parentheses that in our earlier attempts (e.g., references 39 and 40) we tried to work with the local nonadiabaticity parameter and were therefore not very successful.]

One may similarly object that the existence and motion of a sharp partial hydrogen ionization region and strong shocks surely make the pulsation highly nonlinear. Again this is certainly true locally, but the overall effect of these phenomena on the pulsation is not very strong.[22,28] In other words, the amplitude saturation mechanism and the modal interaction are not dominated by these phenomena.

In FIGURE 2 we reproduce stretches of the temporal behavior of the outer radius for a sequence of Population II Cepheid (W Virginis) models with stellar parameters $M = 0.6M_\odot$, $L = 500L_\odot$, $X = 0.7$, and $Z = 0.002$ for the sequence D of Kovács and Buchler.[37] The effective temperature that is indicated in the figures varies from $T_{\rm eff} = 5475$ K to 5200 K. The Hopf bifurcation, which destabilizes the fundamental normal mode, is at a temperature slightly higher than 6200 K. The model sequence displays a metamorphosis from regular periodic to increasingly irregular behavior. From a mere visual inspection of the temporal behavior, however, it is very difficult to understand what is happening, and we have to resort to a different type of analysis of the data. We will show that the evidence is overwhelming that we are witnessing a textbook Feigenbaum route[41,42] from regular behavior to chaos through a period-doubling cascade.

In FIGURE 3 we exhibit the *first return maps*[35,41,42] for the models displayed in FIGURE 2. The return maps are constructed with the successive maximal values of the stellar radius. One clearly witnesses period one, period two, period four, and period eight limit cycles, followed by a two-band and a one-band chaotic cycle, respectively.

FIGURE 4 displays the three-dimensional *phase-space reconstructions*[43,44] for some models of sequence D of reference 37, obtained again from the temporal behavior of the stellar radius. The axes represent $[R(t), R(t + \Delta t), R(t + 2\Delta t)]$. For good visual display Δt is typically 40 percent of the period (or near period). These pictures suggest that the attractor that underlies the irregular pulsations can be embedded in

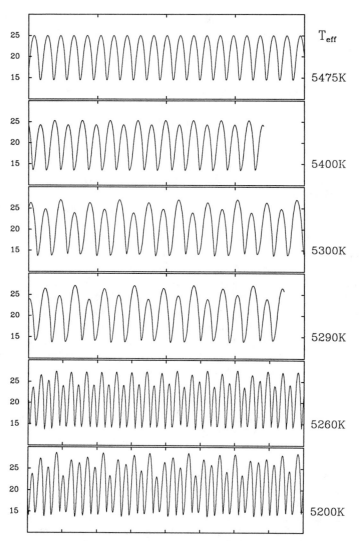

FIGURE 2. Stretches of temporal behavior of the outer radius in [10^{11} cm] for models from a sequence of W Virginis models of mass $0.6M_\odot$, luminosity $500L_\odot$, and Population II composition ($X = 0.745, Z = 0.005$). Note a change of scale in the time axis for the last two models.

a three-dimensional submanifold of phase space. The low dimension of the attractor compared to that of the full phase space of the discretized model ($= 180$, for 60 zones with a radius, a velocity, and a thermal variable each) implies that very few degrees of freedom are involved in the nonlinear behavior and that the attractor is chaotic.

Computations of the correlation dimension from our time-sequences add circumstantial evidence that the dimension of our chaotic attractor is less than three.[37]

In FIGURE 5 we display a power spectrum of the temporal behavior of the radius for the last two, chaotic models shown in FIGURES 3 and 4. It is the largest peak that corresponds to the frequency of oscillation of the fundamental mode. In the top panel the first subharmonic is strong and has its origin in the two-band structure of the attractor (cf. FIG. 4). A closer inspection of FIGURE 4 shows that there is still a

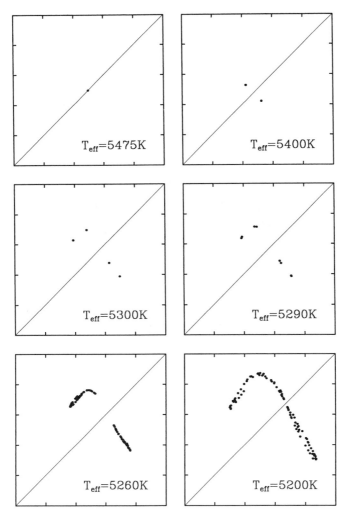

FIGURE 3. First return maps constructed with the successive maxima of the stellar radius for the same sequence as in FIGURE 2.

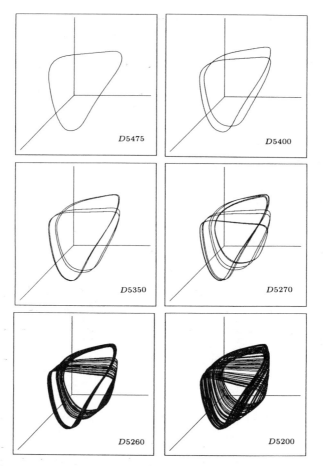

FIGURE 4. Three-dimensional phase-space reconstruction for models of the sequence.

small remnant of the four-band structure that is responsible for the clearly visible sub-subharmonic in FIGURE 5. In the bottom attractor the bands have almost completely merged and the subharmonic structure shows up as tall grass in the Fourier spectrum.

Because these numerical hydrodynamic computations are costly we cannot perform them over a sufficiently dense grid in control space. We therefore have to rely on experience gained from much simpler, more amenable systems. It turns out that the results we have obtained show a remarkable resemblance to the behavior of the Rössler oscillator (e.g., fig. 12.4 in reference 42, or fig. 2 in reference 43). This oscillator is described by the simplest possible set of three equations that have only one nonlinear coupling term,

$$\frac{dx}{dt} = -y - z \tag{1a}$$

$$\frac{dy}{dt} = x + ay \tag{1b}$$

$$\frac{dz}{dt} = b + z(x - c) \tag{1c}$$

and yet behaves in a very complicated way. With parameter values $b = 2$ and $c = 4$, and with the control parameter a ranging from $a = 0.35$ to 0.398 the oscillator undergoes a period-doubling cascade to chaos that displays the same type of banded structure as our phase-space reconstruction (FIG. 4).

For stellar models with effective temperatures lower than the ones shown (cf. fig. 12 of reference 37) the topology of the attractor seems to become somewhat more complicated than the Rössler band, although it still seems to be embeddable in three dimensions.

The return maps for such three-dimensional attractors are of course two-dimensional. For most well-known attractors that are embeddable in three dimensions, however, these return mappings are so dissipative that, in effect, they are almost one-dimensional. This is nicely seen in fig. 12.3 of Thompson[42] for the Rössler attractor and in our return maps in FIGURE 3.

The results that we have presented and their analysis leave little doubt that a

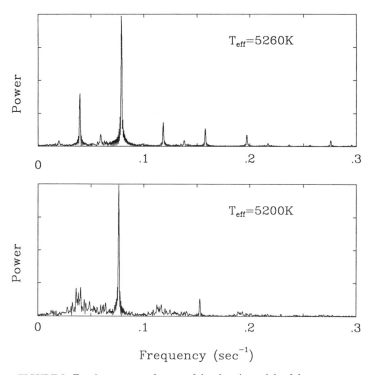

FIGURE 5. Fourier spectrum for two of the chaotic models of the sequence.

chaotic attractor develops in the dynamic of the stellar models and that it is responsible for the erratic temporal behavior.

THE MECHANISM BEHIND PERIOD DOUBLING

It has already been mentioned that a limit cycle that is created in a Hopf bifurcation lives in a two-dimensional submanifold of phase space. This is schematically illustrated in FIGURE 6, where, for simplicity, the shape of the manifold has been taken to be independent of the control parameter. The static state of the system that is stable up to the Hopf bifurcation is described by a point. Because of the uniqueness of the trajectories in phase space, this limit cycle cannot period double unless it leaves this two-dimensional manifold. In the following we address the

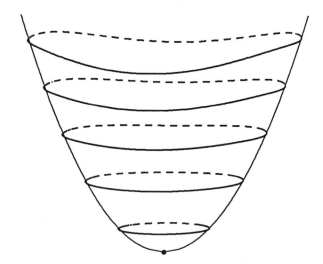

FIGURE 6. Schematic behavior of the limit cycle.

mechanism by which the dimension of the submanifold gets increased so as to permit period doubling. In the spirit of our modal description of the pulsation this (these) additional dimension(s) must come from the eigenvector(s) of an additional mode, whether thermal or vibrational, which somehow becomes destabilized.

Before proceeding we recall elements of Floquet analysis.[45] Consider a fundamental limit cycle, that is, a fundamental periodic nonlinear pulsation that is a solution of the hydrodynamic equations (which, in our discretized approximation, consists of a set of 180 first-order coupled equations). There are therefore 180 linearly independent perturbations $\delta_k(0)$ that can be applied at some initial time $t = 0$. The time evolution of these perturbations over a whole period P_0 of the limit cycle transforms them into $\delta_k(P_0)$. The *Floquet matrix* **F** relates the final to the initial perturbations

$$\{\cdots \delta_k(0) \cdots\} = \mathbf{F} \cdot \{\cdots \delta_k(P_0) \cdots\}. \tag{2}$$

The matrix $\mathbf{F}$ is diagonalizable and its eigenvalues F_k are called *Floquet coefficients*. It is also customary to introduce the Floquet exponents λ_k and phases ϕ_k

$$F_k = \exp(\lambda_k + i\phi_k). \tag{3}$$

Near the fundamental blue edge of a mode, say the fundamental mode 0, the Floquet exponents and phases are closely related to the linear eigenvalues of the static model, basically because perturbing the static model is equivalent to perturbing the limit cycle of infinitesimal amplitude. Thus,

$$F_k = \exp(\lambda_k + i\phi_k) \simeq \exp((\kappa_k + i\omega_k)P_0) \tag{4}$$

and

$$\lambda_k \simeq \kappa_k P_0 \tag{5}$$

$$\phi_k \simeq \omega_k P_0 = 2\frac{P_0}{P_k} \equiv 2\pi\, P_{0k}. \tag{6}$$

Away from the Hopf bifurcation the Floquet coefficients differ more and more from the linear eigenvalues. By carefully following their evolution it is, however, possible to keep track of their modal significance.

As the effective temperature is changed the internal structure of the models changes. As a result all the other stellar properties gradually change as well, including the linear periods of oscillation. In particular, the period ratios P_{0k} evolve and a series of internal resonances are encountered along a given model sequence. For integer resonances, the ratio $P_{0k} = n$, $n \in N$, and one therefore expects the Floquet phase ϕ_k to vanish (mod 2π), whereas at a half-integer resonance $P_{0k} = \frac{1}{2}(n + 1)$, and the phase should be equal to π. In reality the situation is a little more complicated. First, nonlinear effects produce a shift in the periods P_0 and P_k, which can move these points away from the linear resonance condition, and second, nonlinear effects also produce a more characteristic behavior in the vicinity of these resonances.[46,47]

We first discuss the pulsational behavior of a weakly dissipative stellar type, namely the classical Cepheids, although they do not undergo the period-doubling cascade to chaos. The reason for this diversion is, first, that for these stars the behavior near a resonance can be studied with simple amplitude equations when the resonant coupling terms are sufficiently strong to give to *phaselock* (synchronization). Second, they display the same basic mechanism that in the more dissipative Population II Cepheids allows a subsequent period-doubling cascade to chaos to occur. With minor simplifications it has recently been shown analytically[46,47] how the nonlinear terms modify the behavior of the Floquet phases and exponents in the vicinity of the resonances.

(a) For a low-order integer resonance the phase can cross 0 or just approach it depending *inter alia* on the relative strength of the resonant coupling terms and of the saturation terms, that is, ultimately on the stellar model parameters. The exponent generally undergoes a wiggle when the resonance is crossed.[47] Since this type of resonance is not responsible for period doubling, we will not discuss it further here.

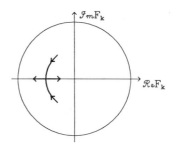

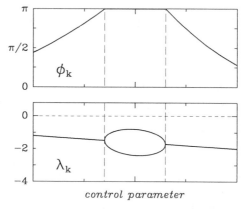

control parameter

FIGURE 7. Schematic behavior of the Floquet phases and exponents in the vicinity of a half-integer resonance.

(b) For a low-order, half-integer resonance, two complex conjugate Floquet coefficients for the resonant vibrational overtone meet on the negative real axis and separate horizontally, then come back together and again move off the real axis back into the complex plane. The corresponding Floquet phases therefore not only necessarily reach π, but exhibit a plateau there while the exponents display a bubble. This behavior is shown schematically in FIGURE 7. The limit cycle is stable as long as the bubble does not pierce 0 (assuming, of course, that all the other eigenmodes are stable). The size and location of the bubbles depends on the model parameters. It can be shown analytically[46,47] that, at the piercing point, the overtone mode is born with an infinitesimal amplitude (*supercritical* bifurcation) and how its amplitude grows away from it. Because the limit cycle now involves a mixture of both modes, the stellar pulsation has a period of $2P_0$ because of the resonance relation and of phase lock. It is important to add that only a single period doubling can occur in weakly dissipative systems and that they undouble back to a regular limit cycle as the control parameter continues to vary.

The importance of the study of this weakly dissipative system is that it shows that at the bifurcation the attractor moves from its original two-dimensional submanifold into a higher dimensional submanifold, and that this bifurcation is the result of a half-integer resonance.

In FIGURE 8 we display the behavior of the Floquet phases and coefficients that have been computed for the least stable Floquet eigenmodes for a $6M_\odot$, $4000L_\odot$ sequence of radiative Cepheid models that are pulsating in the fundamental mode. The identification of the modes has been started at the Hopf bifurcation on the left of the diagram and then followed to lower T_{eff}. Bubbles in the exponents and concomitant plateaux in the phases are clearly visible. Only the lowest four vibrational (complex) Floquet eigenmodes have been shown. All the other modes are

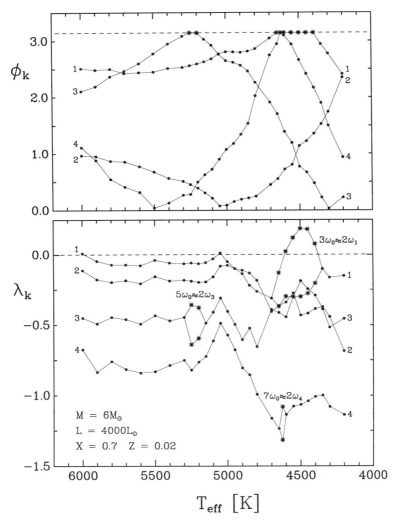

FIGURE 8. Floquet phases and exponents for a classical Cepheid sequence.

more strongly damped and most of them fall below the range shown for λ_k. We mention in passing that the reason why only the lowest vibrational overtones are involved can be very simply understood from the amplitude equations.[46,47] The two smaller bubbles occur for negative λ_3 and λ_4, and therefore have no direct bearing on the limit cycle. (They can only have a transient effect in giving an oscillatory approach toward the limit cycle.) The larger bubble, on the other hand, gives rise to a positive λ_1 over a range of models near 4500 K, and thus destabilizes the limit cycle by adding an admixture of the first overtone and increasing its period to $2P_0$ because of the half-integer resonance and of phaselock.

The linear stability analysis of the sequence shows the following internal resonances among the modes of interest.

$$3\omega_0 = \omega_4 \quad \text{at} \quad 5435 \text{ K,}$$

$$5\omega_0 = 2\omega_3 \quad \text{at} \quad 5235 \text{ K,}$$

$$2\omega_0 = \omega_2 \quad \text{at} \quad 5000 \text{ K,}$$

$$3\omega_0 = 2\omega_1 \quad \text{at} \quad 4510 \text{ K,}$$

$$7\omega_0 = 2\omega_4 \quad \text{at} \quad 4360 \text{ K,}$$

$$3\omega_0 = \omega_3 \quad \text{at} \quad 4100 \text{ K,}$$

The half-integer resonances are indicated in FIGURE 8 next to the bubble with which they are associated.

There is little doubt that the half-integer resonances are responsible for the bubbles and that integer ones give rise to the behavior near zero phase. [We mention in parentheses that it is the $2\omega_0 = \omega_2$ resonance that is at the origin of the so-called Hertzsprung bump progression[28,29] (cf. Moskalik et al. in this volume).] This period doubling, which we find in some Cepheid models, is interesting, but it has not been observed. It should, however, be noted that since nobody has been aware of the possibility of such behavior in long-period Cepheids, it may simply have been overlooked. On the other hand, it is also quite possible that convection, which has been neglected, could have the effect of lowering the bubble below the stability line. Further work on period doubling in Cepheids is in progress.

We now return to the Population II Cepheids for which $\kappa_k P_k \simeq 0.1$–0.5, and which are therefore much more dissipative than the classical Cepheids and RR Lyrae. We show the results for the sequence D of reference 37, which has already been discussed. In this sequence the validity of the amplitude equation formalism breaks down, but some of its predictions remain qualitatively true, as we shall see.

FIGURE 9 displays the behavior of the Floquet phases and exponents for the sequence of W Virginis models whose metamorphosis from regular to chaotic pulsations was discussed earlier. The behavior of these quantities is much less regular, presumably because of the more violent motion of these models and the loss of some accuracy. Nevertheless, several bubbles are clearly visible. The identification of the modal characteristics of the exponents is still possible, although considerably more difficult. Finally, because of the large differences between the linear and the nonlinear periods it is more difficult to associate a specific resonance with a given bubble. The most likely identifications of the relevant two half-integer resonances are indicated inside the bubbles in the figure. The first period doubling, which occurs when the large bubble crosses zero near 5400 K, is associated with the resonance $5\omega_0 \simeq 2\omega_2$ between the fundamental limit cycle and the second overtone. The bubble does not close within the computed range of T_{eff}.

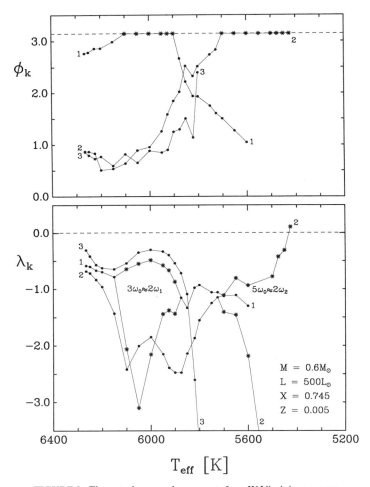

FIGURE 9. Floquet phases and exponents for a W Virginis sequence.

While the first period doubling is still associated with a half-integer resonance, as in the mildly dissipative classical Cepheid models, the subsequent period doublings and chaos cannot be understood on the basis of the amplitude equation formalism. In order to get a better understanding of this strongly nonlinear and dissipative behavior, a system of two coupled oscillators has been studied.[46] These oscillators have been chosen to mimic as closely as possible the coupling between the two resonant modes as encountered in the stellar models, and the results, indeed, bear a striking resemblance to the stellar behavior. In particular, this study has shown that the successive period-doubling boundaries and the curve of onset of chaos form a set of embedded concave parabola-like curves in the parameter plane of dissipation and detuning (the analogues of L and of T_{eff}, respectively). (This is somewhat of an oversimplification, since folds can occur in the bifurcation diagram.) Depending therefore on the path in this parameter space, only a few period doublings might occur and be followed by a return to regular behavior via successive undoublings, without ever becoming chaotic. It is this result from the study of the model oscillators that has led us to compute an additional sequence of stellar models with a somewhat lower luminosity (lesser dissipation).[46,47] As suggested by the model oscillators, this model sequence indeed proceeds to a period two, then a period four cycle, and finally returns to period two and ultimately period one. We therefore feel that on the basis of the amplitude equation formalism, on the one hand, and the model oscillator study, on the other, the resonance mechanism for period doubling is well understood in our W Virginis models.

As an historic note it is interesting that in the study of a W Virginis model Christy[13] suggested in 1966 that a 3:2 resonance might be responsible for producing period two behavior, because in his model he found a linear period ratio of 0.66.

In addition to the period-doubling cascade another well-known, textbook route to chaos, namely a *tangent bifurcation,* has been found in model sequences with higher luminosity to mass ratios.[36,37,48,49] In FIGURE 10 we display the temporal behavior of the outer radius for selected models of sequence A of reference 37 whose parameters are $0.4M_{\odot}$, $400L_{\odot}$, $X = 0.65$, and $Z = 0.005$. The models show stretches of almost regular behavior that are interrupted by bursts of chaotic behavior. The intervals of regular behavior become increasingly shorter and less regular as T_{eff} is lowered. On the other hand, the model with $T_{eff} = 6000$ K (not shown) remains

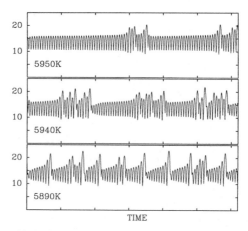

FIGURE 10. Temporal behavior of the outer radius in $[10^{11}$ cm] for models of a sequence with $0.4M_{\odot}$ and $400L_{\odot}$

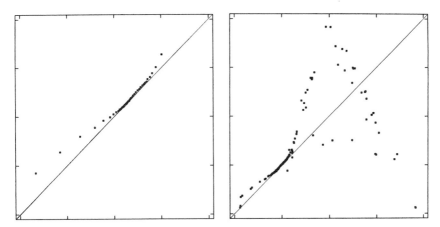

FIGURE 11. First return map for the model with $T_{eff} = 5950$ K of the sequence in FIGURE 10.

regular at all times. As before, the mere visual inspection of the temporal behavior is a poor guide to understanding the nature of the behavior.

The right-hand side of FIGURE 11 shows the first return map constructed from the temporal behavior of the outer radius of the model with $T_{eff} = 4950$ K. The map on the left is a blowup of the region of near tangency where the model gets "trapped" into an almost regular behavior. After it manages to leave this channel, it jumps around the existing chaotic attractor, as seen in the full return map on the right, until a sufficiently large excursion reinjects it into the channel. This behavior is characteristic of a tangent bifurcation.[41,42]

We have seen that irregular behavior arises along well-known routes in radiative models in which convective heat transport has been totally ignored. On the one hand, the question arises whether, in fact, convection can destroy the bifurcation structure that we have found. On the other hand, the neglect of convection is not justifiable, at least for the cooler models, since during part of the pulsation cycle the ratio of convective to total flux approaches unity in some parts of the envelope. To study the effects of convective heat transport, a standard mixing length model for time-dependent convection[50,51] has been included in our hydrodynamic code.[52] The purpose of this study has not been to produce an accurate description of heat transfer, but rather to estimate the sensitivity of the models to some form of convection. Exploratory computations suggest[52] that the whole bifurcation structure, viz., the period doublings and the tangent bifurcation, survive in the presence of time-dependent convection, but get shifted to slightly higher luminosities. This is a comforting result, but more work is necessary to study the effect of time-dependent convection on the bifurcation structure.

The chaotic behavior that we have discussed here is relatively well behaved in the sense that one can still talk about some "period," although it may fluctuate considerably (cf. FIGS. 2, 4, 5). The types of variable stars that display this type of behavior are some W Virginis, of the RV Tau, and possibly of the semiregular stars. The irregularly varying stars in which no such "periodicity" is apparent are all classified as irregular variables. A subgroup of these is characterized by intermittency, that is, the pulsations occasionally stop for a while and then start up again. It has been suggested that the near-dynamical instability (which leads to a very low frequency of the fundamental mode) may give rise to intermittency.[53] In a nutshell, the basic idea is that a resonance couples the fundamental to the first overtone mode and leads to an alternating excitation of the two modes. Since the fundamental frequency of oscillation is very small, there is no observable pulsation when it is this mode's turn to be excited, but merely a slow expansion or contraction. The pulsational timescale is that of the first overtone. This idea has not yet been tested with hydrocodes, partly because of numerical difficulties, but also because the uncertainties of time-dependent convection, which plays a very important role in these objects.

CONCLUSIONS

Sequences of stellar models have been shown to undergo two well-established routes to chaos, viz., a period-doubling cascade and a tangent bifurcation, depending on the stellar parameters. It has further been established that it is the destabilization of an overtone through a resonance of *half-integer* type, viz., $5\omega_0 \simeq 2\omega_2$, which is responsible for an increase in the dimension of the attractor that ultimately allows subsequent period doublings and chaos to occur.

From an astrophysical point of view it is interesting that irregular variability occurs so naturally in the dynamic of the star, even in the absence of convection. Our modeling indicates that there exists a *chaotic blue edge* in stellar parameter space. The occurrence of chaos has been found to be very robust; not only do other hydrocodes produce it as well, but it also survives when time-dependent convection is included. We therefore believe that there is now good theoretical evidence that deterministic chaos is the probable cause of the irregular variability of the W Virginis stars, and possibly also of the RV Tau and the semiregular stars. Of course this work is only the first step in this direction, and more detailed, and especially more accurate numerical hydrodynamical modeling will be necessary.[30] As far as observational evidence is concerned, some attempts have been made to analyze data for the presence of chaos,[54,55] but the paucity and, generally, the unsuitability of the data have prevented any definite conclusions to date.

ACKNOWLEDGMENTS

I wish to thank my collaborators Géza Kovács, Marie-Jo Goupil, Pawel Moskalik, and Ami Glasner for the fun we had during the course of this work.

REFERENCES

1. WHEATLY, C. W., G. W. SCOTT & A. MIGLIORI. 1986. Los Alamos Sci. **4:** 1.
2. PERDANG, J. 1985. *In* Chaos in Astrophysics, J. R. Buchler, J. Perdang, and E. A. Spiegel, Eds.: 11. NATO ASI Series **C191.** Reidel. Dordrecht, The Netherlands.
3. COX, J. P. 1980. Theory of Stellar Pulsations. Princeton University Press. Princeton, N.J.

4. CHRISTY, R. 1966. Rev. Mod. Phys. **36:** 555.
5. COX, J. P. 1980. Space Sci. Rev. **27:** 389.
6. KOVÁCS, G. 1989. *In* The Numerical Modelling of Nonlinear Stellar Pulsations; Problems and Prospects, J. R. Buchler, Ed.: 53. Kluwer. Dordrecht, The Netherlands.
7. DEUPREE, R. G. & S. W. HODSON. 1976. Astrophys. J. **208:** 426.
8. MANTEGAZZA, L. 1985. Astron. Astrophys. **151:** 270.
9. DUONG-VAN, M., ED. 1987. Chaos. North-Holland. Amsterdam, The Netherlands.
10. LANDOLT, H. & R. BORNSTEIN. 1950. Landolt-Bornstein Tables. Springer-Verlag. Berlin/New York.
11. WALLERSTEIN, G. & A. N. COX. 1984. Publ. Astron. Soc. Pac. **96:** 677.
12. SPARROW, C. 1982. The Lorenz Equations: Bifurcations, Chaos and Stange Attractors. Springer-Verlag. New York/Berlin.
13. CHRISTY, R. F. 1966. Astrophys. J. **145:** 337.
14. DAVIS, C. G. 1972. Astrophys. J. **172:** 419.
15. ———. 1972. Astrophys. J. **187:** 175.
16. FADEYEV, YU. A. 1984. Astrophys. Space Sci. **100:** 426.
17. FADEYEV, YU. A. & A. B. FOKIN. 1985. Astrophys. Space Sci. **111:** 355.
18. BRIDGER, A. 1985. *In* Cepheids: Theory and Observations, B. F. Madore, Ed.: 246. Cambridge University Press. Cambridge, England.
19. WORRELL, J. K. 1986. Mon. Not. R. Astron. Soc. **223:** 782.
20. TAKEUTI, M., M. NAKATA & T. AIKAWA. 1985. Sendai Astron. Rep. No. 275:182.
21. BUCHLER, J. R. & G. KOVÁCS. 1986. Astrophys. J. **308:** 661.
22. ———. 1987. Astrophys. J. **318:** 232.
23. BUCHLER, J. R. & M.-J. GOUPIL. 1984. Astrophys. J. **279:** 394.
24. BUCHLER, J. R. 1985. *In* Chaos in Astrophysics, J. R. Buchler, J. Perdang, and E. A. Spiegel, Eds.: 137. NATO ASI Series **C191**. Reidel. Dordrecht, The Netherlands.
25. ———. 1988. *In* Multimode Stellar Pulsations, G. Kovács, L. Szabados, and B. Szeidl, Eds.: 71. Kultura. Budapest.
26. KLAPP, J., M.-J. GOUPIL & J. R. BUCHLER. 1985. Astrophys. J. **296:** 514.
27. BUCHLER, J. R. & G. KOVÁCS. 1986. Astrophys. J. **303:** 749.
28. KOVÁCS, G. & J. R. BUCHLER. 1989. Astrophys. J. **346:** 898.
29. BUCHLER, J. R., P. MOSKALIK & G. KOVÁCS. 1990. Astrophys. J. **351:** 617.
30. BUCHLER, J. R. 1989. *In* The Numerical Modelling of Nonlinear Stellar Pulsations; Problems and Prospects, J. R. Buchler, Ed.: 1. Kluwer. Dordrecht, The Netherlands.
31. STELLINGWERF, R. S. 1974. Astrophys. J. **192:** 139.
32. ———. 1975. Astrophys. J. **195:** 441.
33. KOVÁCS, G. & J. R. BUCHLER. 1988. Astrophys. J. **324:** 1026.
34. FRALEY, G. S. 1968. Astrophys. Space Sci. **2:** 96.
35. BUCHLER, J. R. & G. KOVÁCS. 1987. Astrophys. J., Lett. **320:** L57.
36. BUCHLER, J. R., M.-J. GOUPIL & G. KOVÁCS. 1987. Phys. Lett. **126:** 177.
37. KOVÁCS, G. & J. R. BUCHLER. 1988. Astrophys. J. **334:** 971.
38. SPIEGEL, E. A. 1985. *In* Chaos in Astrophysics, J. R. Buchler, J. Perdang, and E. A. Spiegel, Eds.: 91. NATO ASI Series **C191**. Reidel. Dordrecht, The Netherlands.
39. BUCHLER, J. R. & O. REGEV. 1981. Astrophys. J. **250:** 769.
40. ———. 1981. Astron. Astrophys. **114:** 188.
41. BERGE, P., Y. POMEAU & C. VIDAL. 1986. Order Within Chaos. Wiley. New York.
42. THOMPSON, J. M. T. & H. B. STEWART. 1986. Nonlinear Dynamics and Chaos. Wiley. Chichester, England.
43. PACKARD, N., J. P. CRUTCHFIELD, J. D. FARMER & R. SHAW. 1980. Phys. Rev. Lett. **45:** 712.
44. ROUX, J.-C., R. H. SIMOYI & H. L. SWINNEY. 1883. Physica **8D:** 257.
45. HARTMAN, P. 1982. Ordinary Differential Equations. Birkhauser. Boston.
46. MOSKALIK, P. & J. R. BUCHLER. 1990. Astrophys. J. **355:** 590.
47. BUCHLER, J. R. & P. MOSKALIK. 1989. *In* The Numerical Modelling of Nonlinear Stellar Pulsations; Problems and Prospects, J. R. Buchler, Ed.: 141. Kluwer. Dordrecht, The Netherlands.
48. AIKAWA, T. 1987. Astrophys. Space Sci. **139:** 281.
49. ———. 1988. Astrophys. Space Sci. **149:** 149.

50. Cox, A. N., R. R. Brownlee & D. D. Eilers. 1966. Astrophys. J. **144:** 1024.
51. Ostlie, D. 1989. *In* The Numerical Modelling of Nonlinear Stellar Pulsations; Problems and Prospects, J. R. Buchler, Ed.: 88. Kluwer. Dordrecht, The Netherlands.
52. Glasner, A. & J. R. Buchler. 1989. *In* The Numerical Modelling of Nonlinear Stellar Pulsations; Problems and Prospects, J. R. Buchler, Ed.: 108. Kluwer. Dordrecht, The Netherlands.
53. Buchler, J. R. & M.- J. Goupil. 1989. Astron. Astrophys. **190:** 137.
54. Moskalik, P. 1989. Unpublished.
55. Kollath, Z. 1990. Mon. Not. R. Astron. Soc. In press.

Classical Cepheid Pulsations[a]

PAWEL MOSKALIK,[b,c] J. ROBERT BUCHLER,[c] AND
GÉZA KOVÁCS[d]

[c]Physics Department
University of Florida
Gainesville, Florida 32611
and
[d]Konkoly Observatory
Budapest, Hungary

INTRODUCTION

Cepheids are classical variable stars that undergo radial pulsations with large amplitudes. In the majority of cases the pulsations are almost exactly periodic. Exceptions are the so-called double-mode Cepheids,[1,2] and the rather odd Cepheid HR7308[3] whose pulsation, although rather regular, shows a long-term amplitude modulation. The Cepheid masses are typically between $6M_\odot$ and $20M_\odot$ and the luminosities between $2400L_\odot$ and $16,000L_\odot$. These variables are evolved stars of Population I (metal rich) that cross the instability strip during blue loops in the Hertzsprung–Russell diagram.

The shape of the Cepheid light and velocity curves is known to undergo a systematic change with period. This phenomenon is commonly referred to as the Hertzsprung or bump progression.[4,5] It manifests itself as the appearance of a secondary maximum, or *bump,* whose position is correlated with the period. The bump occurs for the first time on the descending branch for stars with periods P near 7 days. As the period of the Cepheids increases, the bump appears increasingly upwards on the light curves. For Cepheids with periods P greater than 10 days, it appears on the ascending branch, then moves down and evanesces.

A dozen years ago Simon and Schmidt[6] conjectured that this progression was caused by a 2:1 resonance between the fundamental mode and the second overtone. Their arguments were based on a comparison of the linear period ratios of Cepheid models, on the one hand, and the numerical hydrodynamic computations of Stobie,[7] on the other. This conjecture has since been put on a firm basis through numerical hydrodynamical calculations,[8,9] as well as through the development and application of the *amplitude equation formalism.*[10–13]

A precise evaluation of the bump progression is difficult with an eyeball inspection, and Simon and coworkers[14,15] introduced a Fourier analysis of the observed light and velocity curves in order to provide a quantitative description. The observations are fitted with a Fourier sum in the form

$$A_0 + \Sigma A_n \sin(n\omega t + \phi_n). \tag{1}$$

[a]This work was supported by the National Science Foundation, the Pittsburgh Supercomputing Center, and the NER Data Center at the University of Florida.
[b]On leave from the Copernicus Astronomical Center, Warsaw, Poland.

37

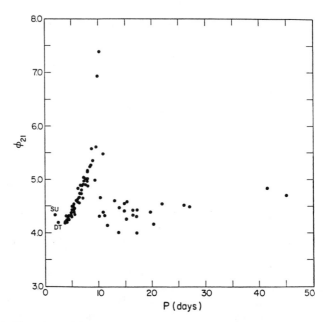

FIGURE 1. The phase difference ϕ_{21} for the V-magnitude curves of classical Cepheids. (Adapted from Simon and Moffett.[16] Courtesy of the *Publications* of the Astronomical Society of the Pacific.) Symbols SU and DT mark suspected first overtone pulsators SU Cassiopeia and DT Cyghus.

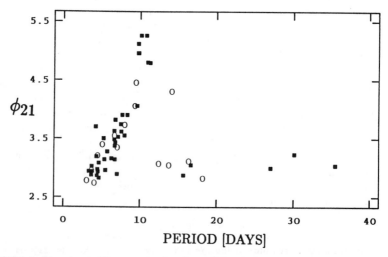

FIGURE 2. The phase difference ϕ_{21} for the radial velocity curves of classical Cepheids. (Adapted from Kovács *et al.*[5]) Symbols $\bigcirc$ denote stars for which the error in ϕ_{21} exceeds 0.25.

The Hertzsprung progression then manifests itself as a characteristic dependence of the derived Fourier parameters $R_{n1} = A_n/A_1$ and $\phi_{n1} = \phi_n - n\phi_1$ on the pulsation period, P. This approach provides a well-defined description because it relies on parameters derived from the whole cycle rather than merely from the position of the bump. It also provides a natural framework within which the model behavior and the observations can be compared.

FIGURE 1 (adapted from Simon and Moffett[16]) and FIGURE 2 (adapted from Kovács *et al.*[5]) present phase ϕ_{21} versus P diagrams for the light and the radial velocity curves, respectively, for galactic Cepheids. (In this paper velocities are in the astrocentric frame.) The two diagrams have a very similar appearance, showing an increase of ϕ_{21} for $P \leq 10$ days followed by a sharp decline (often described as a break, although there is no evidence of a real discontinuity[13]) at 10 days and a relatively flat and featureless tail for $P \geq 10$ days. The larger scatter in the diagram for the radial velocity phases is primarily due to the larger observational errors of Cepheid velocity curves.[5]

SURVEY OF CLASSICAL CEPHEID MODELS

The purpose of this work has been to elucidate the effects of the resonance and to reproduce the bump progression as well as possible, not only qualitatively but also quantitatively. As mentioned in the Introduction, by bump progression we mean the characteristic progression of the Fourier coefficients. In order to optimize our understanding of the systematics of the pulsational behavior our approach has been to compute one-parameter families, or *sequences*, of models, each with constant M, L and chemical composition, and with T_{eff} as the control parameter along the sequences. This choice for the sequences mimics to some extent the Cepheid evolutionary tracks that are horizontal in the H–R diagram. (For the reader, unfamiliar with pulsation theory, we recall that the nuclear evolution timescale is much longer than the pulsational timescale, so that it is generally possible to decouple the pulsational and evolutionary behavior; cf. however, the paper of Lebovitz in this volume.) Our sequences then represent "snapshots" of the pulsational behavior at different stages of evolution.

The methodology of our survey consists of the following four steps.

(a) For each model in a given sequence we first construct the static equilibrium. Following the usual procedure we limit ourselves to the envelope that alone takes part in the pulsation.

(b) A linear normal mode analysis (LNA) of the model is performed that yields the linear stability properties, the periods, and therefore the location of the resonances.

(c) The static model could be perturbed with the velocity component of the linear fundamental mode eigenvector, and its temporal evolution could be followed with a hydrocode. This would generally require the integration through a great number of cycles and would be very time-consuming and expensive, especially if an accuracy of 10^{-4} or better in periodicity is desired. Instead, we directly converge the dynamic behavior to its periodic nonlinear pulsation (*limit cycle*).[17,18] This procedure also guarantees that we are not just witnessing a long-lived transient. Additionally, the Floquet matrix,[19] which is basically a by-product of the relaxation procedure, yields invaluable information about the stability of the limit cycle. (Actually, the method is powerful enough even to converge to mildly unstable limit cycles whose instability shows up as Floquet coefficients greater than unity in modulus.)

We stress that because of the relatively crude meshsize demanded by affordable hydro computations it is essential to use a continuous algorithm for generating the numerical mesh (e.g., with anchors at crucial points) so as to obtain smooth results that can be compared along sequences. It is equally important that the linear analysis and hydrocodes use the same zoning, the same difference scheme, and the same physics (equation of state, an analytical fit to the opacity) to ensure compatibility between linear and nonlinear results.[20]

(d) Finally a Fourier analysis of the light and radius variations of the outermost zone is performed on the converged limit cycles. The coefficients for the velocity curves are then readily obtained from those of the radius variations (because $v = dR/dt$).

RESULTS

We have computed the nonlinear pulsational behavior of a number of classical Cepheid sequences that span a wide range of masses and luminosities. For all

TABLE 1. Stellar Parameters of Cepheid Sequences

Sequence	Symbol	Mass	Luminosity
A	+	4	2500
B	*	5	3000
C	○	5	4000
D	□	6	4000
E	△	8	7500

NOTE: $X = 0.70$; $Z = 0.02$.

computed models the chemical composition has been assumed to be $X = 0.70$, $Z = 0.02$. Observation suggests that the bump Cepheids are fundamental mode pulsators and, in consequence, we have limited ourselves to this type of pulsation. Some of the sequences were merely computed to check the sensitivity of our results to numerical details or to opacities, or they had a more extreme luminosity for a given mass. The other, *standard,* sequences that will be discussed here have a moderate range of luminosities for a given mass even though the range of masses is wide (cf. TABLE 1, which also indicates the symbols that are used in the figures to label the sequences). The complete description of the results of the survey is given in Buchler et al.[9]

FIGURE 3, on the right, adapted from a recent analysis of Kovács et al.,[5] displays the observed (astrocentric) radial velocity curves versus time (t/period). The curves are labeled by their stellar origin and are ordered by period. The progression of the bump is clearly visible. We note that the bump does not "move" from descending to

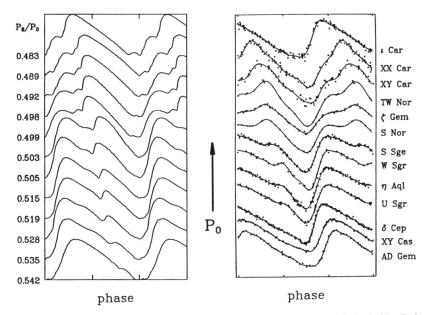

FIGURE 3. Bump progression for the radial velocity curves of classical Cepheids. *Right:* Observations together with the Fourier fits. (Adapted from Kovács *et al.*[5].) *Left:* Theoretical velocity curves for models of sequence *D*.

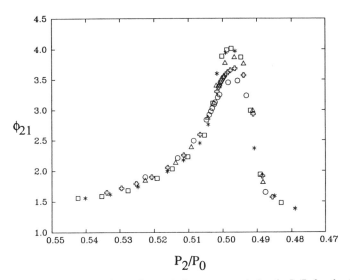

FIGURE 4. Fourier phases ϕ_{21} for radius variation versus period ratio P_2/P_0 for the Cepheid model sequences (cf. TABLE 1 for symbols).

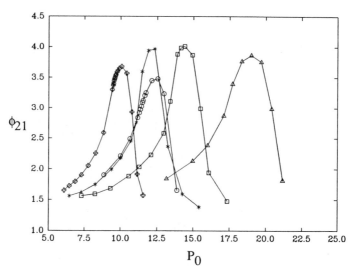

FIGURE 5. Fourier phases ϕ_{21} (radius variation) versus pulsation period P_0, in days, for the Cepheid model sequences (cf. TABLE 1 for symbols).

ascending branch near 10 days, but rather that the primary and secondary maxima switch roles there. The theoretical velocity curves for the models of our sequence *B* are displayed on the left of FIGURE 3. The behavior of the bump is qualitatively well reproduced, although the theoretical bumps are somewhat too pronounced. This can

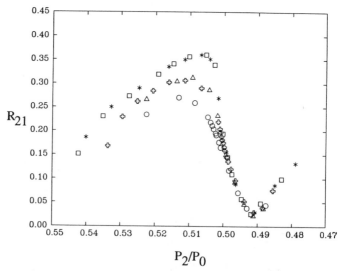

FIGURE 6. Fourier amplitude ratios R_{21} (radius variation) versus period ratio P_2/P_0 for the Cepheid model sequences (cf. TABLE 1 for symbols).

be due to a poor treatment of radiative transfer in the outer zones or it may be due to the fact that we plot the velocity for the outermost zone, whereas the metallic lines used to get observational curves are formed in a deeper, and in addition, non-Lagrangean region. If we plot a deeper zone, the theoretical bumps become less

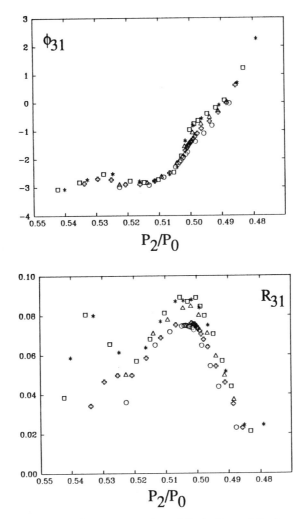

FIGURE 7. Third-order Fourier parameters ϕ_{31} and R_{31} (radius variation) versus period ratio P_2/P_0 for the Cepheid model sequences (cf. TABLE 1 for symbols).

pronounced and resemble the observed ones more closely. We mention that this erroneous accentuation of the bumps has also been found with other hydrocodes.

The most remarkable result of our survey is the very tight correlation between ϕ_{21} for the radius (and radial velocity) variations and the period ratio $P_{20} = P_2/P_0$, which

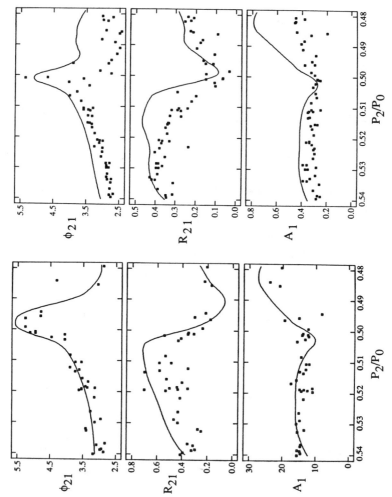

FIGURE 8. Comparison of model sequence B with classical Cepheid observations. *Left:* Fourier parameters for the light curves. *Top, middle,* and *bottom* frames display ϕ_{21}, R_{21}, and A_1 (in [km/s] or in magnitudes), respectively, plotted versus period ratio P_2/P_0. See text for the transformation from P_0 to P_2/P_0. *Right:* Fourier parameters for the radial velocity curves.

is shown in FIGURE 4. Despite the fact that the mass varies from $4M_\odot$ to $8M_\odot$ and the luminosity from $2500L_\odot$ to $7500L_\odot$, we have an almost universal progression of this Fourier parameter with the distance to the resonance center, measured by P_{20}. The ascending and descending branches are both virtually scatterless. The only visible scatter occurs around the maximum. When the same results are plotted versus P_0, however, the correlation is completely lost, as can be seen in FIGURE 5. This does not come as a surprise, since the transformation from P_{20} to P_0 depends strongly on the mass, and the plotted sequences have different masses.

A tight correlation with the period ratio P_{20} is also clearly visible for R_{21} and the higher order Fourier coefficients shown in FIGURES 6 and 7. It is a little less strong for the amplitude ratios, but is still remarkable. Similar results are found for the Fourier coefficients of the light curves.[9] The strong correlation of all the Fourier parameters with P_{20}, but not with P_0 clearly demonstrates that the 2:1 resonance is the dominant effect in the shaping of the pulsations of the bump Cepheids.

In FIGURE 8 we compared our theoretical results with the observations through the intermediary of Fourier analysis. Solid lines represent lowest order Fourier coefficients for model sequence B plotted versus P_{20} together with the Fourier coefficients for the observed Cepheid light and radial velocity curves. The observational coefficients for the radial velocity are taken from Kovács *et al.*[5] and for the light curves from Simon and Moffett.[16] Since P_{20} is not an observable quantity, we need to know the transformation from period ratio to period in order to make this graph. This transformation can be found theoretically, but the calculations give results that are systematically in disagreement with observations and stellar evolution computations. (This problem is commonly referred to as *mass discrepancy.*[21]) We have therefore proceeded as follows. We assume that the transformation has the form

$$P_{20} = CP_0^{-\alpha} \tag{2}$$

and that the resonance center is located at 10 days (where we see a "break" in ϕ_{21} progression). The remaining of the two free parameters C and α is then determined by a least square fit between theoretical and observed values of ϕ_{21} for the velocity. We thus stretch and shift the abscissae, but leave the vertical coordinates unaffected. These same C and α parameters are then used for all the other Fourier coefficients graphs.

FIGURE 8 shows that the agreement for the Fourier coefficients A_1, R_{21}, and ϕ_{21} of the velocity curves is very good. The ratio R_{21} is slightly too high for the plotted sequence B. This is not a serious problem, however, because another sequence, for example, A, gives lower values of R_{21} (cf. FIG. 6) while preserving the quality of the fit for ϕ_{21} and A_1. Now turning our attention to the light curves, we find that the fit of the Fourier coefficients is not so good, and shows systematical discrepancies, especially for ϕ_{21} and A_1 for the long period stars (low values of P_{20}). We nevertheless still reproduce all the qualitative features of the progression. The deterioration of the agreement of theory and observations for light curves is probably a reflection of the poor treatment of radiative transfer and shocks in the outer zones of the models. This problem affects all existing numerical codes to some extent, and as a result the theoretical light curves should be considered less reliable than the velocity curves.

DESCRIPTION WITH AMPLITUDE EQUATIONS

The tight correlation of the Fourier coefficients with P_{20} that is seen in FIGURES 4, 6, and 7 is not the only reason for believing that the $2\omega_0 \simeq \omega_2$ resonance is the main factor in the shaping of the Cepheid oscillations. With the help of the amplitude equation formalism it has become possible to understand the underlying mechanism analytically. This formalism assumes that the behavior of the system is controlled by a small set of dominant or "principal" modes that are either linearly unstable or only slightly damped[10,22,23] (see also Buchler in this volume). The temporal behavior of the amplitudes for these modes is determined by a set of ordinary differential equations (the amplitude equations), and the behavior of the whole system can be expressed in terms of these amplitudes and the linear normal eigenmodes. In our case, the appropriate set of amplitude equations involves the linearly unstable fundamental mode (F) and the resonant, linearly stable second overtone (O_2), and has the form[12]

$$\dot{a} = \sigma_0 a + \Pi_0 a^* c + Q_0 |a|^2 a + T_{02} |c|^2 a \tag{3a}$$

$$\dot{c} = \sigma_2 c + \Pi_2 a^2 + Q_2 |c|^2 c + T_{20} |a|^2 c, \tag{3b}$$

where a and c are the complex amplitudes of F and O_2, respectively, and σ_0 and σ_2 are the linear eigenvalues of these modes $\sigma = i\omega + \kappa$. The quadratic terms describe the 2:1 resonant coupling and the cubic terms describe the nonlinear saturation of the instability mechanism. Higher order terms in the amplitude equations have been found to be unimportant.[24] Since from an astrophysical point of view we are only interested in limit-cycle oscillations (attractors), we look for the asymptotic solution of the preceding amplitude equations. These equations can be simplified with the introduction of amplitudes and phases, $a = A \exp(i\psi_0)$ and $c = C \exp(i\psi_2)$. Limit cycles correspond to phase-locked solutions in which the amplitudes A and C, as well as the phase difference, $\psi_2 - 2\psi_0$, are constant in time.[12]

The behavior of the system is described by

$$|z> = (a|0> + c|2> + \text{c.c.}) + \text{h.o.t.}, \tag{4}$$

where $|z>$ is the "state vector" for the whole star (radius, velocity, and temperature in all zones of the model), and $|0>$ and $|2>$ are linear F and O_2 eigenvectors.[12,22] In the case of the surface radius variations the preceding expression becomes[13]

$$\frac{\delta R}{R_*} = h_{20} |a|^2 + h'_{20} |c|^2$$

$$+ a + h_{21} a^* c + \text{c.c.}$$

$$+ c + h_{22} a^2 + \text{c.c.}$$

$$+ \text{h.o.t.} \tag{5}$$

The consecutive rows of Eq. 5 group terms whose temporal dependence is *constant*, $\exp(i\omega t)$, and $\exp(2i\omega t)$, respectively. The amplitude equation formalism naturally gives a description of the pulsation in the form of a Fourier expansion.

In order to predict the Fourier coefficients for a given stellar model with the amplitude equations we need to know the nonlinear coupling coefficients in Eqs. 3 as well as the expansion coefficients in Eq. 5. In principle, they can be calculated *ab initio,* that is, from the model structure and the linear eigenvectors,[10] but in practice this is a very difficult numerical task. Instead, we therefore have opted to treat the expansion coefficients h_{ij} as well as the coefficients in Eqs. 3, scaled by P_2, as constant parameters along each sequence of models to be determined from fit to hydrodynamical results. Specifically, we have performed a nonlinear least square fit of the solutions of the amplitude equations to the Fourier parameters of the given sequence, treating all the constant (nonlinearity) coefficients as free parameters of the fit (the eigenvalues σ_i are taken from normal mode analysis). Details of this procedure are described in Kovács and Buchler.[13]

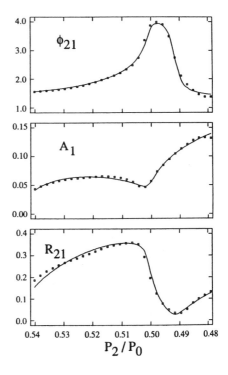

FIGURE 9. Fit of the low-order Fourier coefficients for the radius variations of the model sequence *B*. The hydrodynamic data are shown by *dots,* and the amplitude equation predictions by *solid lines.*

In FIGURE 9 we show the resulting fit for model sequence *B*. The dots represent the Fourier coefficients of the radius variations of the computed hydro models, whereas the solid lines represent those of the solutions of the amplitude equations with coefficients obtained from the fit. The agreement is remarkably good considering that all the nonlinear coefficients are assumed constant along the sequence. We note that for the other sequences a similarly good fit can be obtained and, importantly, that the values of the fitted coefficients are close for all sequences.

We conclude that the nonlinear pulsations of these models and the bump progression, in particular, are well captured by amplitude equations that have been truncated at the lowest order saturation terms. The amplitude equation formalism allows an analytical understanding of the role of the 2:1 resonance and provides new physical insight into this mechanism, and it is the only existing theory capable of doing it. The bump appears through the nonlinear resonant excitation of the otherwise linearly stable second overtone.[11,12] Stated differently the linearly unstable fundamental mode saps energy from the thermal reservoir, this energy gets transferred to the second overtone through the nonliner coupling, and finally, the linearly stable overtone returns this energy to the thermal reservoir.

WIDTH OF THE INSTABILITY STRIP

Our hydro results show that the ϕ_{21} phase for the radius variation (or equivalently for the radial velocity variation) correlates very tightly with P_{20}, but not with P_0 and, to a very good approximation, it can be considered to be the function of P_{20} alone (at least on the ascending branch of the progression). This is true despite the fact that our sequences explore a broad range of M and L (a factor of 2 in M, a factor of 3 in L). Observations do not yield P_{20}, and the observational ϕ_{21} is plotted versus P_0 instead. Nevertheless, for both light and velocity curves a good correlation with P_0 is found (especially for the magnitude m, where observational errors are not dominating as in the case of the radial velocity). We now exploit the consequences of these two facts.

The observed ϕ_{21} (P_0) is seen to have some measurable dispersion at P_0 = const. This dispersion can be used to determine the width of the instability strip. In FIGURE 10 we present a schematic H–R diagram in which we have marked the instability strip, a line of constant period P_0 and lines of constant period ratios P_{20}. The evolutionary mass–luminosity relation has been taken into account, and consequently the lines with P_{20} = const. are not parallel to the lines of constant P_0. Stars with a given pulsation period thus cover some range of period ratios. Since ϕ_{21} depends on P_{20}, *inter alia*, we necessarily see a spread of ϕ_{21} for a specified period. It is easy to notice from this picture that the range of P_{20} (and thus the spread of ϕ_{21}) depends directly on the width of the instability strip. This effect can be estimated quantitatively with the help of simple relations given by evolution and pulsation theories. We start with the formula

$$\log P_0 = 0.83 \log L - 0.66 \log M - 3.45 \log T_{\text{eff}} + \text{const.}, \qquad (6)$$

which represents a fit to the results of a linear normal mode analysis (LNA).[25] In this equation, as in all the following, we express the luminosity and mass in solar units, P_0 in days, and T_{eff} in [K]. (The symbol *log* throughout represents the decimal logarithm.) Stellar evolution theory provides an empirical mass–luminosity relation

$$\log L = \gamma \log M + \text{const.}, \qquad (7)$$

where for γ we adopt the recently found value of 3.5 for the second (and longest) crossing of the instability strip.[26] We also make the realistic, but simplifying assumption that the instability strip is delimited by straight lines in the $\log L$ versus $\log T_{\text{eff}}$ diagram

$$\log L = -\chi \log T_{\text{eff}} + \text{const.} \qquad (8)$$

For χ we adopt the value of 23.5, which was derived for the blue edge from linear pulsational model calculations.[27] We also make the assumption here that both edges of the strip have the same value of χ, that is, that they are parallel. Combining Eqs. 6, 7, and 8, we find the equation for the line of constant period, and the luminosities at which this line is crossing the red and the blue edges, respectively. Thus, we obtain the range of luminosities for all stars with given P_0 that are within the strip,

$$\Delta \log L = 4.38 \, \Delta \log T_{\text{eff}}, \qquad (9)$$

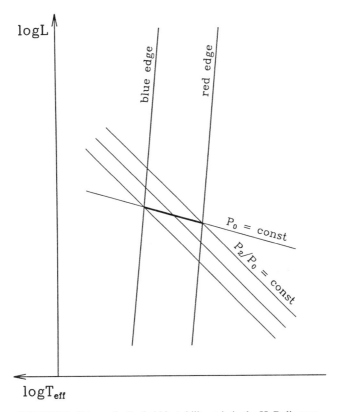

FIGURE 10. Schematic Cepheid instability strip in the H–R diagram.

where $\Delta \log T_{\text{eff}}$ is the (logarithmic) width of the strip measured at constant luminosity. According to the mass–luminosity relation, this range of luminosities corresponds to the range of masses

$$\Delta \log M = 1.25 \Delta \log T_{\text{eff}}. \qquad (10)$$

Since P_{20} depends on the stellar mass, this range corresponds to a certain range of period ratios. According to the LNA calculations P_{20} can be regarded as a function of

M and P_0 alone, in first approximation, which we write in the form

$$\log P_{20} = \theta \log M - \mu \log P_0 + \text{const.} \tag{11}$$

From a least square fit of this formula to our results we find $\theta = 0.138$. A combination of Eqs. 10 and 11 yields the range of period ratios at fixed period:

$$\Delta \log P_{20} = 0.173 \Delta \log T_{\text{eff}}. \tag{12}$$

Together with the assumption that $\phi_{21} = \phi_{21}\,(P_{20})$ this leads to an estimation of the dispersion for ϕ_{21} at constant period, namely

$$\Delta \phi_{21} = 0.173 \Delta \log T_{\text{eff}} \left| \frac{d\phi_{21}}{d \log P_{20}} \right|. \tag{13}$$

Because ϕ_{21} can be obtained relatively easily observationally, the preceding expression permits us to estimate the width of the instability strip $\Delta \log T_{\text{eff}}$. We stress that the value of $\Delta \log T_{\text{eff}}$ obtained that way is only an upper limit because we have not taken into account some other effects that increase the dispersion of $\phi_{21}\,(P_0)$. The most important among them are: observational errors in ϕ_{21}; the dispersion of chemical composition within the studied Cepheid sample (all our standard models have the same composition); and the sensitivity of the theoretical ϕ_{21} to M and L (negligible for radial velocity, but not for light curves).

The preceding method allows us to measure the width of the instability strip independently of a luminosity determination and color-temperature calibration, which are always a major cause of uncertainties in the traditional estimation; instead we rely on *evolution and pulsation theories* that also have uncertainties, but that can be eliminated, at least in principle.

Lest the reader be misled we need to point out that in the current state of theory the use of formula Eq. 11 is the weakest point of the whole procedure, since we know that the theoretical $P_{20}(M, P_0)$ relation suffers from the so-called "mass discrepancy problem." In other words, the LNA calculations predict values of P_{20} that are systematically in error for given stellar parameters. We do not use the whole relation here, however, but only its sensitivity to M. Thus, we assume that the problem is "hidden" in the constant of Eq. 11 and that the coefficient θ is not affected. This temporary uncertainty will disappear with the eventual resolution of the mass discrepancy problem.

Our theoretical velocity curves show much better overall agreement with the observations (cf. FIG. 8), and it is thus preferable to apply the preceding method to the ϕ_{21} phases of the radial velocity curves. From FIGURE 2 we can measure $\Delta \phi_{21} \simeq 0.65$ at $P_0 = 7$ days, whereas from the theoretical plot of FIGURE 4 we find $d\phi_{21}/d \log P_{20} \simeq -33.0$ at $P_{20} = 0.52$. (According to the transformation of Eq. 2, $P_0 = 7$ days corresponds to $P_{20} = 0.52$.) Substitution of these two numbers into Eq. 13 yields $\Delta \log T_{\text{eff}} = 0.114$ or $\Delta T_{\text{eff}} = 1450$ K, if we assume that the blue edge is located at $\log T_{\text{eff}} = 3.8$. This result is much larger than any of the traditional observational estimations[28-30] that result in values between 650 K and 1000 K. The reason for our larger estimate is that the ϕ_{21} of the velocity curves are plagued by relatively large observational errors that actually exceed the "natural" dispersion of its progression with P_0 (cf. FIG. 6 in reference 5). The technology exists for obtaining velocity curves with an accuracy that makes a determination of the width of the instability strip possible. Such a determination just has to await the collection of a sufficiently large sample of high-quality Cepheid velocity curves.

In the meantime we are reduced to using the ϕ_{21} phases derived from the observed light curves, which are known very accurately. From FIGURE 1 we find $\Delta\phi_{21} \simeq 0.35$ (at $P_0 = 7$ days), which, combined with $d\phi_{21}/d \log P_{20} \simeq -35.5$ (at $P_{20} = 0.52$), gives $\Delta \log T_{\text{eff}} = 0.057$ or $\Delta T_{\text{eff}} = 775$ K. We can try to improve this estimation by taking into account the dependence of the theoretical ϕ_{21} on M and L. This dependence has been seen to be practically negligible for the radial velocity ϕ_{21}, but to become more significant for the light curve ϕ_{21}. With this change Eq. 13 takes the form

$$\Delta\phi_{21} = 0.173\Delta \log T_{\text{eff}} \left| \frac{\partial\phi_{21}}{\partial \log P_{20}} + \frac{1}{\theta} \frac{\partial\phi_{21}}{\partial \log M} + \frac{\gamma}{\theta} \frac{\partial\phi_{21}}{\partial \log L} \right|. \qquad (14)$$

Substituting the above values for $\Delta\phi_{21}$ and $\partial\phi_{21}/\partial \log P_{20}$ and the derivatives of ϕ_{21} with respect to M and L we find $\Delta \log T_{\text{eff}} \simeq 0.032$, which corresponds to $\Delta T_{\text{eff}} \simeq 450$ K. This value is significantly below any traditional estimation of ΔT_{eff}, which suggests that something has to be wrong with the coefficients we are using in our equations. It is easy to check that a change in χ or γ provides essentially no help. On the other hand, $\Delta \log T_{\text{eff}}$ is very sensitive to the values of $\partial\phi_{21}/\partial \log M$ and of $\partial\phi_{21}/\partial \log L$. These latter could be seriously in error because our pulsation code, as most other ones, treats radiative transfer rather poorly, and consequently our computed light curves could be inaccurate. Another potential source of uncertainty is the coefficient θ, which comes from the $P_{20}(M,P_0)$ relation that is plagued by the mass discrepancy problem. One of the *ad hoc* suggestions for removing the mass discrepancy has been to increase the opacities.[31,32] This can shift the position of the resonance in such a way that its center will occur at $P_0 = 10$ days for accepted evolutionary masses. Just out of curiosity we have performed the linear analysis for one sequence of models with opacities altered *à la* Andreasen.[32] We find that a reduction of θ by a factor of 3 can easily be achieved. This would bring the estimated instability strip width up to $\Delta \log T_{\text{eff}} \simeq 0.052$, or $\Delta T_{\text{eff}} \simeq 715$ K. Before one draws any definite conclusions about the value of θ, however, one would be advised to first get the estimate of $\Delta \log T_{\text{eff}}$ on the basis of the ϕ_{21} for the radial velocity curves. As discussed before, this estimate does not suffer from the uncertainty associated with the dependence of ϕ_{21} on M and L.

CONCLUSIONS

We have carried out an extensive survey of the radiative classical Cepheid models with the goal of studying the relation between the $2\omega_0 \simeq \omega_2$ resonance and the shape of the light and radial velocity curves. The models have been grouped into one-parameter sequences, with fixed mass, luminosity, and chemical composition, and with T_{eff} as the (only) variable parameter along the sequence. This approach allows us to follow the effects of the resonance in the most systematic way. For all the models the exact periodic solutions (limit cycles) have been computed to an accuracy of better than 10^{-4} and the Fourier parameters have been calculated for the radial velocity and light curves.

Our computations show that the Fourier coefficients obtained for sequences with a moderate luminosity to mass relation exhibit a remarkable uniformity when plotted versus the linear period ratio P_2/P_0. This quasi universality is particularly striking in the case of ϕ_{21} and ϕ_{31} for the velocity variations, but persists for all other coefficients. The tight correlation, however, is totally lost when the Fourier coefficients are plotted as a function of the pulsation period itself. This result is a manifestation of the dominant role played by the resonance between the fundamental mode and the second overtone in the shaping of the nonlinear Cepheid pulsations.

The effects of the resonance on the progression of the Fourier coefficients can be understood analytically with the help of the amplitude equation formalism. Within this formalism it is found that the dynamical behavior of the star is essentially determined by the behavior of the fundamental mode and the second overtone, and the amplitudes of these modes are governed by a simple set of two ordinary differential equations. We have demonstrated that the preceding approach (with some simplifications) allows a remarkably good quantitative reproduction of the variations of the low-order Fourier coefficients along the model sequences. This again confirms that the resonance controls the Cepheid pulsations.

The universal relation between the ϕ_{21} Fourier parameters for the velocity variations and the period ratio P_2/P_0 can be exploited to estimate the width of the Cepheid instability strip. The method does not depend on the observational determination of the absolute luminosities and the effective temperatures of the Cepheids, but relies instead on pulsation and evolution theories. The only required observable quantities are ϕ_{21} coefficients that can be obtained relatively easily and that are not biased by any calibration errors. Whereas this method is potentially very capable, its practical application has to wait until more accurate Cepheid radial velocity curves become available.

REFERENCES

1. BALONA, L. A. 1985. In Cepheids: Theory and Observations, B. F. Madore, Ed.: 17–29. Cambridge University Press. Cambridge, England.
2. SZABADOS, L. 1988. In Multimode Stellar Pulsations, G. Kovács, L. Szabados, and B. Szeidl, Eds.: 1–12. Kultura. Budapest.
3. BURKI, G., E. G. SCHMIDT, A. ARELLANO FERRO, J. D. FERNIE, D. SASSELOV, N. R. SIMON, J. R. PERCY & L. SZABADOS. 1986. Astron Astrophys. **168:** 139–146.
4. HERTZSPRUNG, E. 1926. Bull. Astron. Inst. Neth. **3:** 115–120.
5. KOVÁCS, G., E. G. KISVARSANY & J. R. BUCHLER. 1990. Astrophys. J. **351:**606–616.
6. SIMON, N. R. & E. SCHMIDT. 1976. Astrophys. J. **205:** 162–164.
7. STOBIE, R. S. 1969. Mon. Not. R. Astron. Soc. **144:** 485–509.
8. SIMON, N. R. & C. G. DAVIS. 1983. Astrophys. J. **266:** 787–793.
9. BUCHLER, J. R., P. MOSKALIK & G. KOVÁCS. 1990. Astrophys. J. **351:** 617–631.
10. BUCHLER, J. R. & M.-J. GOUPIL. 1984. Astrophys. J. **279:** 394–400.
11. KLAPP, J., M.-J. GOUPIL & J. R. BUCHLER. 1985. Astrophys. J. **296:** 514–528.
12. BUCHLER, J. R. & G. KOVÁCS. 1986. Astrophys. J. **303:** 749–765.
13. KOVÁCS, G. & J. R. BUCHLER. 1990. Astrophys. J. **346:** 898–905.
14. SIMON, N. R. & A. S. LEE. 1981. Astrophys. J. **248:** 291–297.
15. SIMON, N. R. & T. J. TEAYS. 1983. Astrophys. J. **265:** 996–998.
16. SIMON, N. R. & T. J. MOFFETT. 1985. Publ. Astron. Soc. Pac. **97:** 1078–1089.
17. STELLINGWERF, R. F. 1974. Astrophys. J. **192:** 139–144.
18. KOVÁCS, G. & J. R. BUCHLER. 1988. Astrophys. J. **324:** 1026–1041.
19. HARTMAN, P. 1982. Ordinary Differential Equations. Birkhauser. Boston.

20. BUCHLER, J. R. 1990. *In* The Numerical Modelling of Stellar Pulsations: Problems and Prospects, J. R. Buchler, Ed. NATO ASI Series **C302**. Kluwer. Dordrecht, The Netherlands.
21. COX, A. N. 1980. Ann. Rev. Astron. Astrophys. **18:** 15–41.
22. BUCHLER, J. R. 1985. *In* Chaos in Astrophysics, J. R. Buchler, J. M. Perdang, and E. A. Spiegel, Eds.: 137–163. NATO ASI Series **C161**. Reidel. Dordrecht, The Netherlands.
23. COULLET, P. H. & E. A. SPIEGEL. 1983. SIAM J. Appl. Math. **43:** 776–821.
24. BUCHLER, J. R. & G. KOVÁCS. 1987. Astrophys. J. **318:** 232–247.
25. IBEN, I. & R. S. TUGGLE. 1972. Astrophys. J. **178:** 441–452.
26. CARSON, T. R. & R. B. STOTHERS. 1984. Astrophys. J. **276:** 593–601.
27. IBEN, I. & R. S. TUGGLE. 1975. Astrophys. J. **197:** 39–54.
28. PEL, J. W. 1980. *In* Current Problems in Stellar Pulsation Instabilities, D. Fischel, J. R. Lesh, and W. M. Sparks, Eds.: 1–23. NASA Tech. Memo 80625.
29. DEUPREE, R. G. 1980. Astrophys. J. **236:** 225–229.
30. SCHMIDT, E. G. 1984. Astrophys. J. **287:** 261–267.
31. SIMON, N. R. 1982. Astrophys. J., Lett. **260:** 87–90.
32. ANDREASEN, G. K. 1988. Astron. Astrophys. **201:** 72–81.

Pulsations of Delta Scuti Stars

ARTHUR N. COX

Theoretical Division
Los Alamos National Laboratory
Los Alamos, New Mexico 87545

INTRODUCTION

In this paper we give a general review of the pulsating δ Scuti variables, including the observed light curves and positions of the stars in the Hertzsprung–Russell diagram. Theoretical interpretations from evolution and pulsation calculations give their masses, radii, luminosities, and even their approximate internal compositions. Then we discuss three models of these stars, and use them to study the nonlinear hydrodynamic behavior of these stars, after which we outline the hydrodynamic equations and the Stellingwerf[1] method for obtaining strictly periodic solutions. We also present the problems of allowing for time-dependent convection and its great sensitivity to temperature and density. Tentative results to date do not show any tendency for amplitudes to grow to large unobserved amplitudes, in disagreement with an earlier suggestion by Stellingwerf.[2] Finally, we find that the very small growth rates of the pulsations may even be too small to be useful in seeking a periodic solution.

The δ Scuti variables are the most common type of variable star in our galaxy except for the white dwarfs. This is because stars in the mass range from just over one $M_{\odot}$ up to at least several $M_{\odot}$ pass through the yellow giant instability strip in the Hertzsprung–Russell diagram as they evolve off the main sequence to the red. Actually, stars up to the maximum main sequence mass also evolve through this region at higher luminosities, but there are so few of them, and they evolve so rapidly to the red, that they are almost unknown. At the higher luminosity, they probably would be called first-instability strip-crossing Cepheids anyway. Such Cepheids are difficult to separate from those that are on the second blueward instability strip crossing that is much slower. Really, the δ Scuti variables are just low-luminosity Cepheids.

Stars of both populations start their homogenous composition evolution on their main sequences, and at ages usually less than 10^9 years they cross the δ Scuti instability strip. They will be observed as radially and nonradially pulsating stars if their helium content in the surface pulsating layers is large enough for their natural pulsation driving to overcome their deep radiative damping.

Mean radii for these stars when they are pulsationally unstable range from about 1.3 to over 5 $R_{\odot}$ During the pulsations the radius changes over a cycle by only a few percent, as contrasted with the much larger pulsation amplitudes of the Cepheids. The radius and light amplitudes are small for these stars, as observed, because the pulsation driving is smaller and more easily saturated relative to the damping, than for the higher luminosity Cepheids.

Even though the masses of these variable stars are not very much larger than that for our sun, the mean luminosities range from 5 to over $50L_\odot$. This makes these stars easily observable in nearby galactic clusters. Indeed, extensive studies of these stars have been done by many including Breger,[3] who has reviewed the pulsations of these stars.

The instability strip in the Hertzsprung–Russell diagram is between about 6900 K and 8100 K. This range is due to the fact that at the hot side of the strip the driving of pulsations from the cyclical variation of the opacity (κ) effect is becoming too shallow to involve enough of the mass, and at the cool side, the time-varying convection is disrupting this κ effect driving.

Models of these stars reveal that there is indeed very strong convection carrying, typically all but a few percent of the luminosity at the maximum. As discussed later, this convection is able to keep up rather well with the changing configuration during the pulsations, and therefore may be able to interfere with the periodic radiation flow blocking caused by the κ effect.

TABLE 1. Theoretical Masses, Radii, and Luminosities for Double-mode δ Scuti Variables

Variable	$\Pi_0(d)$	Π_1/Π_0	$T_e(K)$	$M_T/M_\odot$	$R_T/R_\odot$	$L_T/L_\odot$	$Q_{0(d)}$
SX Phe*	0.05496	0.778	7850	1.1 ± 0.1	1.3 ± 0.1	5 ± 1	0.0325
CY Aqr†	0.06104	0.744:	7930	1.4 ± 0.1	1.7 ± 0.1	10 ± 1	0.0326
ZZ Mic	0.0654	0.763	7500	1.4 ± 0.1	1.8 ± 0.1	10 ± 1	0.0327
AE U Ma	0.08602	0.773	7500	1.6 ± 0.1	2.2 ± 0.3	14 ± 4	0.0328
RV Ari	0.09313	0.773	7500	1.6 ± 0.1	2.3 ± 0.3	14 ± 4	0.0328
BP Peg	0.10954	0.772	7500	1.8 ± 0.1	3.0 ± 0.3	25 ± 5	0.0328
AI Vel	0.11157	0.773	7620	1.8 ± 0.1	2.9 ± 0.3	25 ± 5	0.0328
V703 Sco	0.14996	0.768	7000	1.9 ± 0.1	3.9 ± 0.5	33 ± 8	0.0329
VX Hya	0.22339	0.773	6980	2.2 ± 0.1	4.8 ± 0.4	48 ± 8	0.0330

*First crossing assumed.
†First or third crossing.
T_e values for SX Phe, CY Aqr, AI Vel, and VX Hya from McNamara and Feltz.[15]
For V703 Sco, T_e is from Jones.[14] For all variables the deep interior composition is $Z + 0.01$ with Y between 0.2 and 0.3, except for SC Phe with $Z = 0.001$ with interior Y between 0.2 and 0.3.

Due to the small size of these stars, their pulsation periods range from just 0.05 to 0.25 day. Only the convection-forced solar oscillations and the white dwarf pulsations have shorter periods for their intrinsic pulsations. Many of these δ Scuti variables display two periods, and often the ratio is between 0.76 and 0.79. This ratio is that expected from the radial fundamental and first overtone pulsation modes. For a few stars the ratio is slightly higher, and it has been suggested that the modes occurring then are the first and second overtone. Helium depletion in the surface layers may, however, change the stellar structure enough to interpret still the modes as the fundamental and first overtone (Cox, McNamara, and Ryan[4]). Stars like 4 CVn, and a few others (Breger et al.[5]), which show five pulsation modes, must have some or even all these modes nonradial with low degree, so that they are observable in the integrated light of the entire stellar disc.

TABLE 1 gives observational data and the theoretically interpreted masses, radii, and luminosities for nine δ Scuti variable stars that have two modes, apparently the

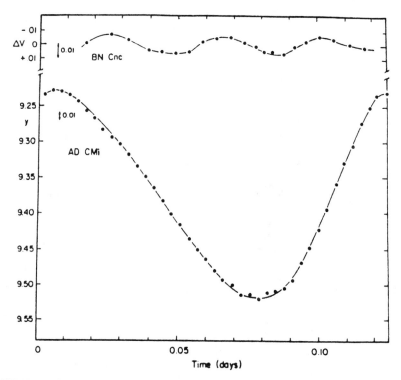

FIGURE 1. Light curves for a typical low-amplitude and a typical high-amplitude δ Scuti variable star. (As presented by Breger.[3] Reproduced by permission.) Light curves of two δ Scuti stars: BN Cnc ($A_v = 0^m014$) and AD CMi ($A_v = 0^m294$). Most δ Scuti stars resemble BN Cnc. However, a continuous range of amplitudes exist from 0^m01 to 0^m8.

fundamental and first overtone. There has been some question as to whether there should be an enhancement of the opacity of the stellar material between temperatures like 150,000 K and 800,000 K that would make theoretical period ratios larger (Andreasen[6]). In the earlier work presented in TABLE 1, these effects have not been included, but the derived masses, radii, and luminosities would not be changed much from those given, because the data depend mostly on theoretical evolutionary track data in the Hertzsprung–Russell diagram.

These data could be derived even more simply for a given variable star by noting the luminosity intersections of the observed effective temperature line and the evolutionary tracks and the luminosity intersections of the constant temperature line with lines of the observed constant period for a series of masses. The stellar luminosity and mass would then be at that intersection luminosity where the mass given by the evolution track is the same as that given by the constant (observed) period line. Actually the exhaustion of the central hydrogen causes a jog in the evolution tracks just after the main sequence stage, and there are sometimes three possibilities for the luminosity at a given effective temperature. The blueward track, however, is very rapid, and it is not likely that we see δ Scuti stars there. Fortunately, the theoretical period ratio for these observed stars at the listed mass, luminosity, and radius is within 0.01 of that observed. One should note how accurately the masses, radii, and luminosities are constrained.

Breger,[3] in his review, plots the light curves for two extreme examples of δ Scuti variables. These are shown in FIGURE 1. Unlike the larger amplitude Cepheids, these stars do not show bumps in their light curves other than those that are the result of multimode superpositions. With such small amplitudes, it is very accurate to compare observed periods with the theoretical predictions, because the linear theory used is not as far from reality as for many other classes of variable stars. The several tenths of a magnitude light variation corresponds to ±10 km/s approximately, and that corresponds to only ± several percent radius variation, typically.

FIGURE 2 gives Breger's[3] observations of δ Scuti variable candidates, and shows the instability strip in the Hetzsprung–Russell diagram. Note that a large fraction of the stars do not pulsate at the minimum observable level of a few thousandths of a magnitude. It could be that the amplitudes are even smaller, or, as I prefer to think, these stars have lost helium by slow diffusive settling from the surface-pulsation driving region near 40,000 K.

FIGURE 3 shows the helium abundance situation as published by Cox, King, and

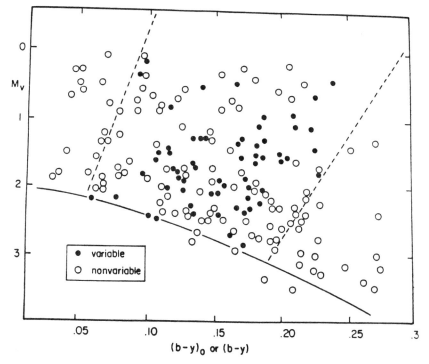

FIGURE 2. Variables and nonvariables are plotted on the Hertzsprung–Russell diagram. (From observations by Breger.[3] Reproduced by permission.) The limits of the variables in temperature and luminosity define the δ Scuti variable instability strip.

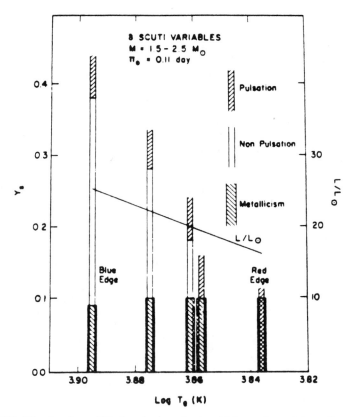

FIGURE 3. The abundance of helium in the pulsation driving regions of the stellar envelope gives pulsation, no pulsation, and metallicism.

Hodson.[7] With advancing age, the helium content (plotted on the left ordinate) decreases by settling downward, unless the star rotates quite rapidly to keep it mixed. The decreased helium, to about 0.1 in Y, eventually causes the helium convection zone to disappear. Then the single well-mixed hydrogen convection zone is considerably more shallow, and radiation levitation processes can work for some metals at much lower densities. Then the surface hydrogen convection zone can quickly become contaminated with metallicism, producing δ Delphini stars or Am stars. The theoretical prediction is then that the simultaneous occurrence of both pulsations and metallicism can be only for the coolest δ Scuti variables. This prediction has been realized recently for HD 1097, where Kurtz (private communication) has found pulsations in an Am star, which Berger[3] reports is just at the red edge of the pulsation instability strip.

Another prediction is that near the red edge, stars do not need much helium to pulsate, and therefore stars there should all be variable. This does not accord with the Breger[3] observations of FIGURE 2, unless the stars are all extremely helium poor and display conspicuous metallicism. Observation of double-mode δ Scuti variables near the red edge should show an occasional anomalous period ratio as predicted by Cox, McNamara, and Ryan[4] for low helium abundance in the surface layers.

On the right ordinate of FIGURE 3 is an indication of the luminosity of the models for a fixed period of 0.11 day. At higher effective temperature the luminosity is higher for fixed period, that is, almost for a fixed radius.

The motivation for our hydrodynamic studies is that Stellingwerf[2] suggested that there is a theoretical "catastrophe" that predicts a very large limiting pulsation amplitude of these stars. The possibility of this giving significant mass loss was discussed by Willson, Bowen, and Struck-Marcell.[8] This then led to calculations (Guzik[9]) of evolution tracks for stellar models for masses just above $1M_\odot$. They indicated that standard theoretical predictions of evolution tracks may need major modifications. Can we, with the latest techniques, predict the observed low pulsation amplitudes, or do we still have the theoretical problem?

The work reported here is an ongoing collaboration with Joyce Guzik, and Dale Ostlie. The goals are to produce a time-dependent convection hydrodynamic program and to apply it to δ Scuti and RR Lyrae variable stars.

TABLE 2. δ Scuti Variable Star Model Parameters

Model	Deep Radiation	Shallow Radiation	Convection
Effective temperature (K)	7263	7263	7800
Radius (10^{11} cm)	2.116	2.116	1.835
Outer shell mass (g)	6×10^{23}	6×10^{23}	2×10^{23}
Outer shell optical thickness	0.06	0.06	0.02
Central ball mass (g)	9.8×10^{32}	2.6×10^{33}	1.18×10^{33}
Central ball mass fraction	0.27	0.73	0.33
Central ball radius fraction	0.08	0.25	0.11
Mixing length ratio	—	—	1.768
H Convection zone minimum T (K)	—	—	7800
H Convection zone maximum T (K)	—	—	26000
Radiative luminosity fraction minimum	—	—	0.022 at 12800 K
He convection zone minimum T (K)	—	—	35000
He convection zone maximum T (K)	—	—	51000
H convection zone minimum q	—	—	5×10^{-11}
H convection zone maximum q	—	—	1.5×10^{-9}
H convection zone minimum x	—	—	0.9979
H convection zone maximum x	—	—	0.9999

NOTE: $1.8\,M_\odot$, $23.2\,L_\odot$, $X = 0.70$, $Z = 0.02$, 60 mass shells.

EQUILIBRIUM MODELS

A homogeneous composition model of $1.8M_\odot$ with composition $X = 0.70$ and $Z = 0.02$ was evolved using the Iben[10] evolution program from the main sequence to the middle of the δ Scuti instability strip at an age of 0.94×10^9 years. This gives a luminosity of $23.2L_\odot$. The evolution was advanced enough to exhaust the hydrogen completely from the inner 8 percent of the mass, and hydrogen is burned slightly all the way out to a temperature of about 5×10^6 K, with 80 percent of the mass interior. In spite of this enhancement of helium at the bottom of our pulsation models by about 0.02 in Y, we assume that the pulsation envelopes are homogeneous with the surface Y of 0.28.

TABLE 2 gives the properties of three models that we have considered for linear theory radial and nonradial pulsations, as well as nonlinear 60 mass shell hydrodynamic calculations. All these studies are nonadiabatic by including the energy conservation equation, with radiation allowed to flow in the diffusion approximation. For the convection model, the standard mixing-length theory was used. To keep the convection as weak as possible for our hydrodynamics, the surface effective temperature for that model was set at 7800 K, rather than the 7263 K for the deep and shallow radiation models.

LINEAR RADIAL AND NONRADIAL PULSATIONS

The 553 mass shell evolution model was reproduced by our model code with 1600 zones to calculate selected radial and nonradial linear periods and growth rates. This model includes the entire mass of the model, and it includes the detailed composition change due to nuclear burning in the central regions. We were able to confirm the general behavior of the growth rates that was previously given by Lee[11] and shown in FIGURE 4(a). FIGURE 4(b) shows the Guzik calculations of the variation of the kinetic energy growth rate (here labeled work) as a function of the angular frequency in units of radians per second (labeled omega). Only the radial and dipole ($l = 1$) p-modes are given.

Lee[11] shows that the variations with frequency are even larger as the l value is increased to 2 and 3. Presumably, these variations are caused by the relative sampling of the driving and damping regions by the modes. We will see these regions later with the much coarser hydrodynamic zones.

The evolution model was also recast into three different 60 mass shell models for the hydrodynamic calculations. TABLE 3 gives the periods, period ratios for adjacent modes, and the growth rates in units of kinetic energy change per period for the first seven radial modes with this coarse zoning. Comparison with FIGURE 4(b) shows that the deep radiative model reproduces the more accurate growth rates well, but the shallow model does not properly sample the fundamental mode damping that is deeper than where it has resolution. It also senses much more damping for the higher order modes, because it has better resolution in the surface layers than the deep model.

The shallow model mode periods given in TABLE 3 are a few percent smaller than for the deep model. This is because the immovable bottom boundary condition is too shallow to represent the true nonzero normal mode motions. Note that the convection model data in TABLE 3 with the smaller radius has shorter periods for all its modes, as expected.

The temperature variation of the opacity is given for the deep radiative model in FIGURE 5. This quantity is the key one for determining the pulsational stability of

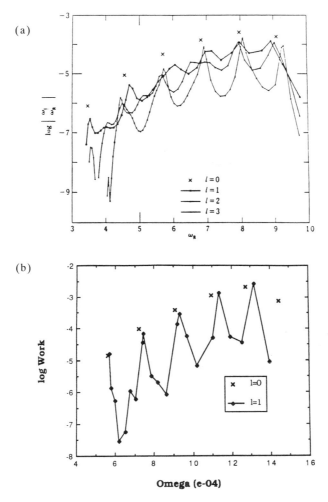

FIGURE 4. 1.8 solar mass δ Scuti star unstable modes. **(a)** The growth rate for radial and nonradial pulsation modes for δ Scuti star models. (Plotted by Lee[11] for the low-order p-modes. Reproduced by permission.) **(b)** The kinetic energy growth rate is plotted for the $l = 0$ and 1 modes for similar models constructed by Guzik (private communication).

stars, because it indicates whether the opacity during the compression stage of a pulsation cycle increases or decreases. If the opacity is higher during compression, radiation is temporarily blocked. Then during the reexpansion phase this radiation can flow easier to increase the internal energy and pressure of the overlying mass layers as they move outward. This phasing gives pulsation driving. If the opacity is decreased during compression, however, radiation can leak away and not be available when the expansion phase occurs. Then this gives radiation damping, which is common in the deep layers of all stars.

Three temperature regions are seen in FIGURE 5 where the opacity either increases with an increase of temperature, that is, during compression, or at least does not decrease very much. Even a negative value of the temperature partial derivative is not necessarily damping, because upon compression the density increases with always an opacity increase from this density effect. The zones near the surface on the right are in the hydrogen and first helium ionization region, with typically the opacity increasing as the twelfth power of the temperature. Near zone 47, the second helium ionization is occurring in the model. Finally, the second helium ionization coupled with the increasing photon energy at about 150,000 K gives the Stellingwerf[12] "bump" near zone 36. All these opacity "bumps" give the well-known κ effect. The low gamma of the material in the hydrogen and helium ionization regions promotes very large compressions and not very large temperature increase there, so that the opacity does indeed increase in the compression stage. The hiding of the luminosity from the γ effect is another process that adds to the driving. Deeper, the

TABLE 3. Radial Mode Periods, Period Ratios, and Growth Rates

Model	Period (day)	Period Ratio	$\Delta KE/KE$ per Period
	Deep Radiative Model		
Fundamental	0.130	0.771	0.0000028
First Overtone	0.100	0.816	0.000049
Second Overtone	0.082	0.839	0.00021
Third Overtone	0.069	0.858	0.00037
Fourth Overtone	0.059	0.876	−0.00022
Fifth Overtone	0.052	0.889	−0.0038
Sixth Overtone	0.046	—	−0.0118
	Shallow Radiative Model		
Fundamental	0.128	0.762	0.0000035
First Overtone	0.098	0.807	0.000066
Second Overtone	0.079	0.832	0.00028
Third Overtone	0.066	0.854	0.00040
Fourth Overtone	0.056	0.873	−0.00100
Fifth Overtone	0.049	0.886	−0.0070
Sixth Overtone	0.044	—	−0.0179
	Convective Model		
Fundamental	0.105	0.769	0.0000070
First Overtone	0.081	0.815	0.000097
Second Overtone	0.066	0.840	0.00053
Third Overtone	0.055	0.857	0.00167
Fourth Overtone	0.047	0.875	0.0040
Fifth Overtone	0.041	0.890	0.0076
Sixth Overtone	0.037	—	0.0096

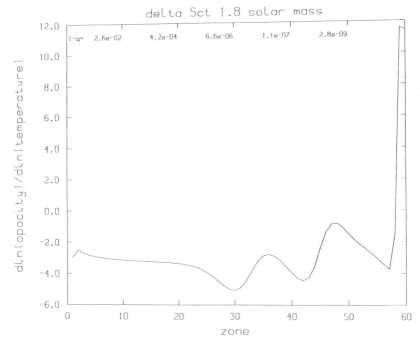

FIGURE 5. The logarithmic derivative of the opacity with respect to temperature is plotted versus zone for the 60 mass shell deep radiative model. Three "bumps" are shown that correspond to pulsation driving regions.

material is "harder" without the ionization occurring, and radiative damping always occurs in stars at temperatures greater than those for the ionization of an abundant element like hydrogen or helium.

FIGURE 6 shows the eigensolution for the work per zone each pulsation cycle as a function of the zone numbers again. One can see that the hydrogen ionization region is driving, but there is not much mass involved to move the entire stellar model. Most of the driving for δ Scuti variables is due to the ionization of helium. There is a nonnegligible amount due to the Stellingwerf[12] bump near zone 31, also. The integral, which one can make by eye in FIGURE 6, is positive, giving pulsation instability for this deep radiative model.

HYDRODYNAMIC METHODS AND RESULTS

The 60 zone models are run in the hydrodynamics program with the equilibrium hydrostatic structure plus the linear δr/r eigenvector superimposed as the velocity structure. All problems have been started with an outward surface velocity of 10 km/s. This velocity amplitude is equivalent to the few tenths of a magnitude light variation that is seen in many δ Scuti variables.

The completely implicit hydrodynamic method used either three or four equations, depending upon whether convection is allowed or is completely ignored in

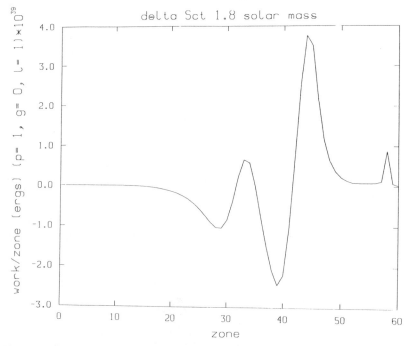

FIGURE 6. The work for each mass shell, each pulsation cycle is plotted versus the zone number. This nonradial p_1 mode is for $l = 1$. The driving is determined by the opacity behavior as well as the mass in each shell and the timescale of the radiation flow compared to the pulsation period.

both the equilibrium model and the hydrodynamics calculation. The equations are the conservation of energy with units of erg/g, the conservation of momentum with units of cm/s², the velocity integral equation with units of cm/s, and the convective velocity equation with again units of cm/s. The form used has all four G_j values equal to zero when the time step, dt, between the previous time p and the final time f is completely converged.

$$G_1 = Ef(i) - Ep(i) + \frac{Pf(i) + Pp(i)}{2}(Vf(i) - Vp(i)) - dt\frac{(sourcef(i) + sourcep(i))}{2}$$

$$+ \frac{dt}{mass1(i)}(\theta(Lf(i) - Lf(i-1)) - (1-\theta)(Lp(i) - Lp(i-1))),$$

$$G_2 = \frac{area(i)}{mass2(i)}\left[\frac{Pf(i+1) + Pp(i+1)}{2} - \frac{Pf(i) + Pp(i)}{2}\right] - g(i) + \frac{(\dot{r}p(i) - \dot{r}f(i))}{dt},$$

$$G_3 = \frac{2}{dt}(rf(i) - rp(i)) - (\dot{r}f(i) + \dot{r}p(i)),$$

$$G_4 = v_c f(i) - velf(i).$$

In these equations for *mass*1 shell-centered and *mass*2 interface-centered, E is the internal energy including turbulent energy; P is pressure including turbulence, turbulent viscosity, and artificial viscosity; V is the specific volume of the shell mass; *source* is the nuclear energy source; θ (usually set to $\frac{2}{3}$) is the off-centering factor on luminosity for stabilization of the implicit method; L is the luminosity including both radiation and convection; *area* is a specially time-centered interface area Fraley;[13] g is the local acceleration of gravity; $\dot{r}$ is the interface velocity, with r the interface position; and $v_c f$, and *velf* are the previous iteration convection velocity and the current convection velocity that is a function of all local variables, respectively.

These equations express the basic concepts of mass, momentum, and energy conservation plus a constraint for the convection luminosity. The mass-conservation equation is not one of the four just given, because it is not a differential equation. Yet is relates the fixed Lagrangian shell mass with density using the time-changing interface radii. The energy equation is the standard first law of thermodynamics. The *PdV* work term allows for the energy to be converted to or from kinetic and potential energy. This role of the *PdV* term can be seen from consideration of the momentum equation. The momentum equation comes in two parts because the pressure gradient produces an acceleration, a second-order time derivative. Thus, the second and third equations are both expressions of the momentum conservation. The convection luminosity equation is a constraint that assures that during the iterations at the f step, the convective velocity at the previous iteration is the same as that given by its analytic expression at the current iteration. Convergence is attained when these equations have their G less than typically 10^{-6} of their largest term.

The bottom boundary condition is an unmovable interface so that the conditions like temperature, density, pressure, and energy interior do not play any role. The luminosity across that bottom interface is the surface luminosity. The surface boundary conditions are that the gas pressure beyond the mesh is zero, while the radiation pressure is the same as in the outermost mass shell. There is no convection luminosity across the surface, and the radiation luminosity is given by the standard Eddington approximation

$$Lf(I) = 4\pi r f(I)^2 \sigma T_e^4$$

with

$$Tf(I)^4 = 3/4 T_e^4(\tau + 2/3)$$

where τ is the Rosseland mean opacity optical thickness of the last mass shell of temperature $Tf(I)$ and specific volume $Vf(I)$. The last interface $mass2(I)$ is taken traditionally 7.7 times the last shell mass for the momentum equation G_2 to roughly account for the exterior chromosphere and coronal mass.

The solution method for these equations is to assume than an extrapolated p-step is a good approximation for the preceding equations. Then the increments to the four independent variables T, r, $\dot{r}$, and v_c are calculated to make the basic hydrodynamic equations become zero. Thus,

$$G_j + \sum_i \frac{\partial G_j}{\partial T_i}\delta T_i + \sum_i \frac{\partial G_j}{\partial r_i}\delta r_i + \sum_i \frac{\partial G_j}{\partial \dot{r}_i}\delta \dot{r}_i + \sum_i \frac{\partial G_j}{\partial v_{c,i}}\delta v_{c,i} = 0.$$

The sums consist of 15 terms for the analytic G_1 derivatives, 19 terms for the analytic G_2 derivatives, 2 terms for the analytic G_3 derivatives, and 11 terms for the analytic G_4 derivatives. Solution of the derivatives matrix for a correction vector for the four independent variables with the residual G_j vector as the right-hand side is done repeatedly with new derivatives calculated each iteration. This Newton–Raphson method results in all $G_j = 0$ at convergence.

This method works extremely well for purely radiative models, but there is considerable trouble when the convection is allowed to transport luminosity. In all cases, it is necessary to limit the corrections to reasonable values so that the solution iterations do not diverge. The convection luminosity is highly nonlinear and very sensitive to the values of the independent variables. Thus, the allowance for complicated and extreme nonlinearity by recalculating the matrix elements each iteration is not necessarily enough to assure convergence. Frequently the time step needs to be reduced to perhaps $\frac{1}{400}$ or less of the pulsation period to allow convergence.

Analysis of convection zones in pulsating stars reveals that there is a good deal of sensitivity of the convection luminosity to the independent variables T and r. The convection luminosity goes as the tenth power of T and as the 10^4 power of r, but only as the tenth power of density.

The convection luminosity is calculated allowing for the time delay that occurs when the convective elements are forced to adjust to the changing pulsation configuration. The convective velocity is interpolated with a three-point Lagrangian method between the velocity at timestep $n - 2$, step $n - 1$, and the current time n, where the current time instantaneous value v^0 is given directly by the mixing-length formulas. For any zone at the interpolated time, t',

$$v_c^n = \frac{(t' - t^{n-1})(t' - t^{n-2})}{(t^n - t^{n-1})(t^n - t^{n-2})} v_c^0 + \frac{(t' - t^n)(t' - t^{n-2})}{(t^{n-1} - t^n)(t^{n-1} - t^{n-2})} v_c^{n-1}$$

$$+ \frac{(t' - t^n)(t' - t^{n-1})}{(t^{n-2} - t^n)(t^{n-2} - t^{n-1})} v_c^{n-2},$$

where

$$t' = t^{n-1} + \tau(t^n - t^{n-1}),$$

with

$$\tau = \frac{v_c^n(t^n - t^{n-1})}{\ell} \, lfac,$$

and τ, the fraction of a mixing length the mean convective eddy can move in one time step, is limited to be between 0 and 1. We have used $lfac$ as unity, but it is available for any adjustment of the lagging.

This convective velocity at the current timestep is checked to see that it has not increased or decreased more than physically possible by the buoyancy or drag forces. This increment is

$$\delta v_c = gdt \left| \frac{\Delta \rho}{\rho} \right| afac = gdtQ \left| \frac{\Delta T}{T} \right| afac$$

where ΔT is the excess temperature above or below the adiabatic gradient between adjacent mass shells, Q is the thermodynamic partial derivative of density with respect to temperature at constant pressure and composition, and *afac,* the acceleration factor, is another adjustable parameter near unity.

There is also spatial averaging for the mean convective element velocity. The velocity used in the mixing-length formula for the convection luminosity is then

$$\bar{v}_{c,i} = a_{i-1}\, v_{c,i-1}^{n-1} + a_{i+1} v_{c,i+1}^{n-1} + \left(1 - a_{i-1} - a_{i+1}\right) v_{c,i}^{n}$$

with $v_{c,i}^{n}$ the lagged convective velocity. The weighting factors for zone k is

$$a_k = \left(1 - \frac{|r_k - r_i|}{\ell}\right) avgfac,$$

subject to the constraint that a_k be between zero and *avgfac,* and *avgfac* being less than or equal to $\frac{1}{3}$.

The turbulent pressure and energy per gram are

$$P_t = \rho \bar{v}_c^2 tfac \qquad \text{and} \qquad E_t = \bar{v}_c^2 tfac.$$

Here *tfac* is taken as $\frac{1}{3}$. Further, the turbulent viscosity pressure, applied both on compression and expansion, depends on the velocity gradient and is

$$Q_t = \frac{\rho \bar{v}_c H_p \Delta \dot{r}}{\Delta r qfac},$$

with H_p being the pressure scale height in the model and *qfac* being of the order of unity.

In addition, there are two other viscosities. Hydrodynamic calculations that produce strong shocks require artificial viscosity to maintain numerical stability. This artificial viscosity is applied only when a zone is being compressed, and it uses the same expression as previously given, with the convection-element velocity replaced with the zone-interface velocity gradient. There is also a linear viscosity used for the deepest mass shell in our hydrodynamics program, and it is calculated as previously, but with the sonic velocity replacing the convective-element velocity. For both of these viscosities, we have canceled the pressure scale height with the interface position difference. We have reduced the coefficient in these viscosities to be sure that the artificial damping is small compared to the natural driving and damping of the stellar pulsations. We also use a modified Stellingwerf[1] procedure of not applying the artificial quadratic viscosity until the relative interface closure velocity exceeds a threshold of 0.02 times the sonic velocity.

The hydrodynamics with convection is made difficult because of the temperature gradient becoming suddenly sub- or superadiabatic. This gives a large change in the convective and total luminosities. When the gradient becomes subadiabatic after being superadiabatic, the convective velocity is not put directly to zero, because physically there will be a drag on the convective elements that is not able to stop the turbulence immediately. The convective luminosity in a subadiabatic region then should be put in as a negative (backwards) value, but we have not developed this procedure yet.

The time behavior of the convection luminosity for the convection model is given in FIGURE 7. At this early stage of calculations we cannot tell whether this convection luminosity is lagging enough to produce pulsation driving. Apparently it does not lead the maximum compression to nullify the periodic luminosity blocking from the κ effect. The convection does reach about 90 percent of the total luminosity each cycle, but for only a brief time. Note that the convection behavior with almost all the luminosity as convection luminosity at the start of the run is greatly modified later by the time lagging and spatial averaging.

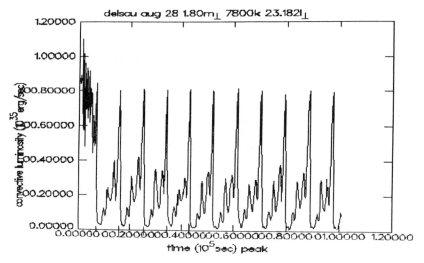

FIGURE 7. The convection luminosity is plotted versus time for the 7800 K model. Note that for a brief time each cycle, the convection carries about 90 percent of the total luminosity.

PERIODIC SOLUTION METHODS

Stellingwerf[1] developed a method for quickly finding a full-amplitude nonlinear solution for stellar pulsations. We have added it to our regular hydrodynamics code so that we can reach limiting amplitude in only a few trial periods instead of calculating the initial-value problem for possibly millions of periods of the mode of interest. Not only does this method yield the desired solutions, but it also can be used to study the stability of the full-amplitude solution against switching to another of the normal modes. Thus, if double-mode behavior occurs, the full-amplitude solutions of the two modes should be unstable against switching to each other.

The method starts by postulating that there is a periodic condition, such that, $\overline{Z}_j^0 - \overline{Z}_j^N = 0$, where Z_j stands for T_j, r_j, $\dot{r}_j$, and v_{cj}, and $\overline{Z}_j$ is the desired periodic solution with complete closure of all variables at the end of the period. Since we want to correct all the independent variables before each trial period, and we also want to adjust the period each trial, we need another equation for this additional quantity $\delta\Pi$. This extra equation is a uniqueness condition $\bar{r}_i^0 - r_i^0 = 0$. In other words, zone i will always be started with the same radius value for all the trial periods and in the final periodic solution. We find that the method works best when this fiducial zone is deep enough to have a temperature at the starting time between 100,000 and 500,000 K.

We use a Newton–Raphson method to converge on the periodic solution:

$$Z_j^0 - Z_j^N + \left(\frac{\partial Z_j^0}{\partial Z_j^0}\right)\delta Z_j^0 - \sum_k \left(\frac{\partial Z_j^N}{\partial Z_k^0}\right)\delta Z_k^0 - \left(\frac{\partial Z_j^N}{\partial\Pi}\right)\delta\Pi = 0,$$

where

$$\overline{Z}_j^0 = Z_j^0 + \delta Z_j^0$$

and

$$\overline{\Pi} = \Pi + \delta\Pi.$$

The method adjusts Z_j^0 and Π to get both the periodicity and the uniqueness. For this latter condition,

$$r_i^0 - r_i^N + \delta r_i^0 - \sum_k \left(\frac{\partial r_i^N}{\partial Z_k^0}\right)\delta Z_k^0 - \frac{\partial r_i^N}{\delta\Pi}\delta\Pi = 0.$$

One can see that the period correction is the fiducial radius closure error divided by the derivative of the fiducial radius with respect to the period

$$\delta\Pi = \frac{r_i^0 - r_i^N}{\left(\partial r_i^N/\partial\Pi\right)},$$

since the postulated correction to r_i is

$$\delta r_i^0 = \sum_k \frac{\partial r_i^0}{\partial Z_k^0}\delta Z_k^0 = 0.$$

With this $\delta\Pi$ known, it is now possible to solve for the δZ_j^0 for all j. The previously given Newton–Raphson equation can be written in the form:

$$\delta Z_j^0 = c_j\delta\Pi + d_j,$$

where the c_j vector carries information about how each zone reacts to the period change, and the d_j vector indicates how good the closure was at the end of the trial period for zone j. These vectors are the result of a matrix solution, where the matrix is the sum of the unity matrix $(\partial Z_j^0/\partial Z_j^0)$ and the Floquet matrix $(\partial Z_j^N/\partial Z_k^0)$. Thus, the matrix here is the Floquet matrix modified by having unity subtracted from each main diagonal element.

The problem now is to see how to relate each independent variable at the end of the trial period to all of them at the beginning, and to relate all variables to a change in the period. This is done by induction during each trial period using always a fixed

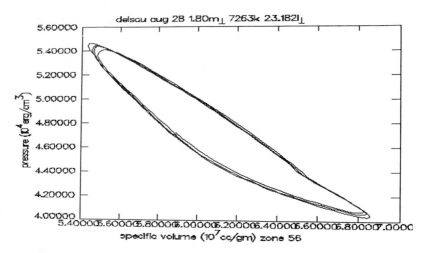

FIGURE 8. Several cycles of the deep radiation model are shown with the trajectory for a zone in the P–V diagram. As the periodic solution is reached, the cycle repeats. This hydrogen driving zone has the looping in the clockwise direction.

set of hydrodynamic timesteps. The equations of energy, momentum, and velocity for zone j in the regular hydrodynamics are

$$G_j^n \left(Z_k^n, Z_k^{n-1}, \Pi \right) = 0,$$

where k denotes all neighboring zones and zone j. Taking partial derivatives of this G_j, we get

$$\delta G_j^n = \left(\frac{\partial G_j^n}{\partial Z_k^n} \frac{\partial Z_k^n}{\partial Z_j^0} + \frac{\partial G_j^n}{\partial Z_k^{n-1}} \frac{\partial Z_k^{n-1}}{\partial Z_j^0} \right) \delta Z_j^0$$

$$+ \left(\frac{\partial G_j^n}{\partial Z_k^n} \frac{\partial Z_k^n}{\partial \Pi} + \frac{\partial G_j^n}{\partial Z_k^{n-1}} \frac{\partial Z_k^{n-1}}{\partial \Pi} + \frac{\partial G_j^n}{\partial \Pi} \right) \delta \Pi$$

$$= 0.$$

Here the multiplications and additions are matrix operations. Note that for this to be zero regardless of the corrections to Z_j and Π,

$$\frac{\partial Z_k^n}{\partial Z_j^0} = - \left(\frac{\partial G_j^n}{\partial Z_k^n} \right)^{-1} \frac{\partial G_j^n}{\partial Z_k^{n-1}} \frac{\partial Z_k^{n-1}}{\partial Z_j^0}$$

and

$$\frac{\partial Z_k^n}{\partial \Pi} = - \left(\frac{\partial G_j^n}{\partial Z_k^n} \right)^{-1} \left(\frac{\partial G_j^n}{\partial \Pi} + \frac{\partial G_j^n}{\partial Z_k^{n-1}} \frac{\partial Z_k^{n-1}}{\partial \Pi} \right).$$

The G_j derivatives are given by analytic formulas in the computing program, and for the derivatives at timestep $n + 1$, they are already known from their use in the implicit hydrodynamics.

The value of $(\partial Z_k^{n-1}/\partial Z_j^0)$ for the first timestep is a matrix with zero everywhere except ones along the main diagonal, and the matrix $(\partial Z_k^{n-1}/\partial \Pi)$ is zero. Each timestep the solution matrix from the previous timestep can be used on the right-hand side to give the updated solution matrix. This induction gives $(\partial Z_k^N/\partial Z_j^0)$ after the fixed N timesteps of each trial period.

This periodic technique has been implemented including the time-dependent convection. The only results to report here, however, are for the purely radiative models. An example of how the looping of a pulsation driving zone appears in the P–V plane is given as FIGURE 8. As the iterations converge, the loops become identical. While convergence in this example is not yet attained to a very small error, it does seem that convergence will occur at this amplitude very close to that observed for these variable stars.

CURRENT CONCLUSIONS

Linear theory studies of δ Scuti variables have reasonably well mapped the behavior of models in both radial and nonradial modes across the instability strip. The current frontier is to understand the nonlinear behavior of these stars, especially when proper allowance for time-dependent convection is made. An interesting question will be why the amplitudes are limited to such small values. Near-main sequence pulsators like the B stars, the δ Scuti stars, and even the sun, are low-amplitude pulsators. Are they still limited in their amplitudes by nonlinear effects as the classical yellow giants are? When our studies are finished, we should be able to answer this question, at least for the δ Scuti stars.

It may be that Stellingwerf[2] did not include in his nonlinear calculations enough deep damping. We expect that our models are deep enough to consider properly this effect, and the low amplitudes that we get may be more nearly correct. When our models are tightly converged, we will be able to comment more completely about this question.

Convection has been implemented in the periodic Stellingwerf[1] method, and runs are being made. The extreme sensitivity of the convection luminosity has not yet been overcome, however.

From early runs for very slowly growing modes, it appears that we may not be able to use the periodic method if the growth rate in kinetic energy is less than 10^{-4} per period. We know this because the Floquet matrix $(\partial Z_k^N/\partial Z_j^0)$, of rank 3 or 4 times the number of Lagrangian mass shells of the model, can be analyzed for its eigensolutions, which are the normal modes of the stellar model. They represent all the ways that the other modes can either grow or decay in the environment of the current hydrodynamic cycle. Among these eigensolutions is one that represents only a phase change of the current solution. That is, it corresponds to the vector ∂Z_j^0 at the trial period start, which is exactly the same as the actual timestep increment of all the independent variables times a real normalization constant. The eigenvalue of this eigensolution should be real and exact unity, but it differs from unity by about one part in 10^4. If the accuracy and stability of the hydrodynamic integration can be improved, though, we may be able to find eigenvalues even closer to unity and reach smaller growth rates. We have already programmed the key parts of the matrix manipulations in double precision to avoid any possible almost exact cancellations.

Our very tentative conclusion is that we see no reason for the large amplitudes that Stellingwerf[2] found in his periodic solutions. Perhaps the deep damping that we have or the time-dependent convection will give us the amplitude limitation that we need.

REFERENCES

1. STELLINGWERF, R. F. 1974. Astrophys. J. **192:** 139.
2. STELLINGWERF, R. F. 1979. Lecture Notes in Physics **125.** H. A. Hill and W. A. Dziembowksi, Eds.: 50. Springer-Verlag. Berlin/New York.
3. BREGER, M. 1979. Publ. Astron. Soc. Pac. **91:** 5.
4. COX, A. N., B. J. MCNAMARA & W. RYAN. 1984. Astrophys. J. **284:** 250.
5. BREGER, M., B. J. MCNAMARA, F. KERSCHBAUM, L. HUANG, S.-Y. JIANG, Z.-H. GUO & E. PORETTI. 1990. Astron. Astrophys. **231:** 56.
6. ANDREASEN, G. K. 1988. Astron Astrophys. **201:** 72.
7. COX, A. N., D. S. KING & S. W. HODSON. 1979. Astrophys. J. **231:** 798.
8. WILLSON, L. A., G. H. BOWEN & C. STRUCK-MARCEL. 1987. Comments Astrophys. **12:** 17.
9. GUZIK, J. A. 1988. Ph.D. Thesis, Iowa State University, Ames.
10. IBEN, I. 1965. Astrophys. J. **141:** 993.
11. LEE, U. 1985. Publ. Astron. Soc. Japan. **37:** 279.
12. STELLINGWERF, R. F. 1978. Astrophys. J. **83:** 1184.
13. FRALEY, G. S. 1968. Astrophys. Space Sci. **2:** 96.
14. JONES, D. H. P. 1975. *In* IAU Symposium No. 67, Variable Stars and Stellar Evolution, V. E. Sherwood and P. Plaut, Ed.: 243. Reidel. Dordrecht, The Netherlands.
15. MCNAMARA, D. H. & K. A. FELTZ, JR. 1976. Publ. Astron. Soc. Pac. **88:** 510.

Bifurcation and Unfolding in Systems with Two Timescales[a]

NORMAN R. LEBOVITZ

Computational and Applied Mathematics Program
Department of Mathematics
University of Chicago
Chicago, Illinois 60637

INTRODUCTION

Many physical problems are formulated in tractable mathematical form by taking advantage of the presence in them of small parameters that can be ignored in a first approximation and provide a means of generating corrections by a perturbation procedure. This is true particularly of astrophysical and geophysical problems wherein the small parameter is the ratio of two widely separated timescales. In this case, exploiting the smallness of the parameter leads to "quasi-steady" equations of motion: some of the dynamical equations are expressed as if the system were in a steady state. An example is standard stellar-evolution theory. The equations representing hydrodynamic motions are written in steady-state form, as are those representing energy conservation; those representing the time rate-of-change of nuclear species are (usually implicitly) retained. If the evolution under consideration takes place during a phase when nuclear transformations do not take place, only the hydrodynamic equations may be regarded to be in a steady state; the energy equation remains in time-dependent form. These reflect the general principle that the slowest process governs the dynamics.

Suppose T_s and T_L are two timescales, short and long, respectively, and let ϵ, their ratio, be a small parameter. If x and λ are variables changing on the fast and slow timescales, respectively, the equations describing the dynamics may be written in either of the following two ways:

$$\epsilon \frac{dx}{dt} = G(x, \lambda), \qquad \frac{d\lambda}{dt} = F(x, \lambda) \tag{1}$$

or

$$\frac{dx}{d\tau} = G(x, \lambda), \qquad \frac{d\lambda}{d\tau} = \epsilon F(x, \lambda). \tag{2}$$

In the first of these, T_L is used as the timescale; in the second, T_s and $t = \epsilon\tau$. For these equations to be general enough to apply to fluid dynamics, they would have to be set in infinite-dimensional spaces, but we will suppose that they are finite-dimensional, that is, that x and λ are vectors of size n and m, respectively. Exploiting the smallness

[a]This work was supported in part by National Science Foundation Grant DMS8903244 with the University of Chicago.

73

of ϵ means setting it equal to zero:

$$0 = G(x, \lambda), \qquad \frac{d\lambda}{dt} = F(x, \lambda) \tag{3}$$

or

$$\frac{dx}{d\tau} = G(x, \lambda), \qquad \frac{d\lambda}{d\tau} = 0. \tag{4}$$

Whereas equations (1) and (2) are equivalent, the limiting equations (3) and (4) are not.

Equation (3) is the "quasi-steady" approximation in that the variable x is treated as if it could be obtained from the steady-state dynamics. One can envision solving Eqs. (3) by first solving the quasi-steady equation $G(x, \lambda) = 0$ for x as function of λ, say $x = X(\lambda)$, and then solving the differential equation for λ obtained by substituting $X(\lambda)$ for x in the second of Eqs. (3). This gives the "reduced" path, and indeed achieves a significant simplification. Equations (4) have a different meaning. In this sytem λ is "frozen" and only the variable x changes. It is precisely the system we would obtain if we wished to study the stability, on the short timescale T_s, of the equilibrium solution $x = X(\lambda)$ for fixed λ. It is furthermore clear heuristically that this picture of quasi-steady evolution only makes sense if these equilibriums are stable on that timescale, and breaks down when that stability is lost.

We are interested in what happens near a point of bifurcation. This is a point (x, λ) at which the steady-state path becomes indeterminate. The function $G(x, \lambda)$ may have more than one zero $X(\lambda)$ (or none at all) at a bifurcation point. If there is more than one path $x = X(\lambda)$, a criterion is needed for deciding which one, if either, the system will choose to follow. If there is none, the motion necessarily takes place on the short timescale once the system passes the bifurcation point. We will restrict consideration to the case when there are two or more solutions in the neighborhood of the bifurcation point. A natural criterion in this case is that the system will choose the stable branch, since it could hardly choose an unstable branch. We wish to investigate the reliability of this criterion. The investigation, and indeed the nature of the dynamics, depends on whether the stability of the equilibriums is asymptotic stability, as prevails for dissipative systems, or nonasymptotic, as prevails for conservative systems. We divide the discussion accordingly.

DISSIPATIVE DYNAMICS

Stability of the equilibriums on the short timescale is the key to the validity of Eqs. (3) for motion on the long timescale. This is easily illustrated by considering the system (1) in the case $n = m = 1$, when x and λ are both scalars. Suppose further that $F(x, \lambda) > 0$ and divide the first equation by the second to get

$$\epsilon \frac{dx}{d\lambda} = \frac{G(x, \lambda)}{F(x, \lambda)}. \tag{5}$$

The path followed by the exact system (2) is determined, as seen in FIGURE 1, by the equilibrium solution of Eq. (5), which is precisely the solution of the quasi-steady Eqs. (3). This conclusion is based on the assumption that for each fixed λ that equilibrium solution is attracting, that is, is asymptotically stable.

This highly plausible behavior is well confirmed mathematically by a theorem originally due to the mathematicians Tikhonov[1] and Levin and Levinson[2] (see references 2 and 3 for more general formulations of the version given here).

THEOREM 1: Let F, G be twice-continuously differentiable in a domain D in R^{n+m}. Let (3) have a solution $x^{(0)}(t), \lambda^{(0)}(t)$ in D for $t_0 \leq t \leq t_1$. For each eigenvalue μ of the Jacobian matrix $G_x(x^{(0)}(t), \lambda^{(0)}(t))$ suppose $Re\mu \leq -\mu_0$ for some $\mu_0 > 0$. Then if $|x(t_0) - x^{(0)}(t_0)|$ is sufficiently small, the solution $x(t, \epsilon), \lambda(t, \epsilon)$ of (1) satisfies the

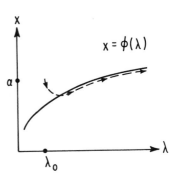

FIGURE 1. The solid curve $x = \phi(\lambda)$ is the family of equilibrium solutions of Eq. (5). The *dashed curve* represents a solution of this equation with initial data (λ_0, a).

following conditions in the limit as $\epsilon \to 0$:

$$x(t, \epsilon) = x^{(0)}(t) + O(\epsilon) \text{ uniformly on } [t_0 + \delta, t_1],$$

where δ is any positive number, and

$$\lambda(t, \epsilon) = \lambda^{(0)}(t) + O(\epsilon) \text{ uniformly on } [t_0, t_1].$$

The restriction that $|x(t_0 - x^{(0)}(t_0)|$ be small is satisfied if the initial point $x(t_0)$ lies in the basin of attraction of the stable equilibrium point $x^{(0)}$ of (4). That $x(t, \epsilon)$ is approximated by $x^{(0)}$ only on an interval excluding the left-hand endpoint t_0, is a reflection of the loss of initial data in going from (1) to (3).

The principal assumption in this theorem regarding the stability of the reduced path is embodied in the eigenvalue condition on the Jacobian matrix G_x. This condition is violated at a bifurcation point: Either one of the eigenvalues of G_x vanishes at such a point (steady bifurcation), or the real part of a complex-conjugate pair vanishes (Hopf bifurcation). We will restrict consideration to the simplest case when only one eigenvalue (or the real part of one pair of complex-conjugate eigenvalues) vanishes at a critical point (x_c, λ_c) (without loss of generality, we take $(x_c, \lambda_c) = (0, 0)$). In this case, there are two kinds of steady bifurcation allowing equilibrium solutions for values of λ in an interval containing λ_c.

One of these, the so-called "transcritical" bifurcation, is illustrated in FIGURE 2 for the case when $n = m = 1$. It is characterized by the intersection of two solution curves of (3), X_1 and X_2, both of which have finite tangents at the intersection. The arrows indicate the direction of the flow for the full system (1), for which ϵ is small and positive. The branches of the curves are labeled S and U for stable and unstable. The dashed curve represents an actual trajectory of (1). In FIGURE 2(a), where the slope of X_1, the curve stable to the left of the bifurcation point, is positive, this acutal trajectory stays close to the stable branches (X_1 to the left and X_2 to the right), as expected. In FIGURE 2(b), where the slope of X_1 is negative, it does not, and it is evident that the quasi-steady equations are unable to describe the dynamics, which takes place on the short timescale T_S. These conclusions can be drawn from consideration of Eq. (5). This behavior is not restricted to the case $n = m = 1$, but is

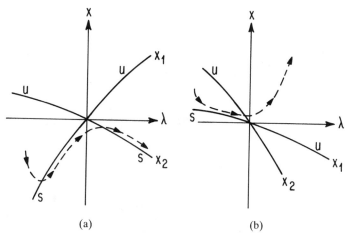

(a)   (b)

FIGURE 2. Transcritical bifurcation. For $\lambda < 0$ the curve X_1 is stable, X_2 unstable; for $\lambda > 0$ this is reversed. Stable branches are marked S. (a) Solutions of Eq. (6) (*dashed curve*) follow stable branches. (b) They do not.

duplicated for arbitrary n and m provided the assumption made previously, that only one eigenvalue of G_x vanish at the bifurcation point, holds. This is also confirmed by a precise theorem like that quoted before, with the difference that in an $\epsilon^{1/2}$-neighborhood of the origin, the approximation is good only to order $\epsilon^{1/2}\log \epsilon$. Since a full statement of this theorem is cumbersome, we refer the reader to reference 3.

The other kind of steady bifurcation we consider is the "pitchfork" bifurcation illustrated in FIGURE 3. We have illustrated the postcritical pitchfork (i.e., the pitchfork occurs for $\lambda > 0$), which is stable (the precritical pitchfork is unstable). With the same conventions as in FIGURE 2, we see that the solution follows stable branches. The figure shows only the case when the slope of the branch that is stable to the left is negative. If it were positive, the acutal trajectory would follow the bottom branch of the pitchfork. This case therefore conforms to expectations. Again

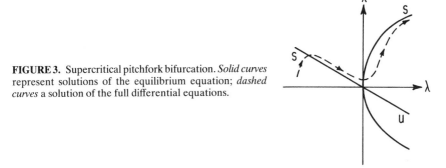

FIGURE 3. Supercritical pitchfork bifurcation. *Solid curves* represent solutions of the equilibrium equation; *dashed curves* a solution of the full differential equations.

a precise theorem is available in the general case of arbitrary n and m, with the difference that in an $\epsilon^{1/2}$-neighborhood of the origin the approximation is good only to order $\epsilon^{1/4}\log \epsilon$. Again we refer the reader elsewhere[4] for details.

The last case we consider is that of Hopf bifurcation, that is, bifurcation from steady to periodic solutions. This occurs when a pair of complex-conjugate eigenvalues $\rho \pm i\nu$ passes from negative to positive ρ as λ passes from negative to positive, whereas ν remains of one sign, and represents the frequency of the small-amplitude oscillations immediately beyond the bifurcation point (FIG. 4). We have again assumed that the periodic branch of solutions is postcritical (and hence stable). In FIGURE 4(a) we indicate the expected behavior of the actual solution. This has been confirmed recently by the Soviet mathematician Neishtadt.[5] But Neishtadt's remarkable result is that this expected behavior does *not* occur if the system (1) is analytic. In that case, the jump to the periodic solution is delayed by a distance of order one, as indicated in FIGURE 4(b). Perhaps equally remarkable is the fact that this delayed instability has been found in numerical computations.[6]

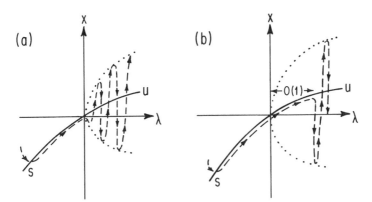

FIGURE 4. Supercritical Hopf bifurcation. The *dotted curve* represents the envelope of the periodic oscillations. (a) The nonanalytic case: oscillations begin after a time approaching zero with ϵ. (b) The analytic case: oscillations begin only after a time of order one.

A delay of the instability in problems of steady bifurcation is also possible, under the following very special condition: suppose the reduced path is an exact solution of the *full* system (1). Then Theorem 1 is modified so that the order of approximation is not ϵ but $e^{-k/\epsilon}$ for some positive constant k. Hence in FIGURES 2 and 3, when the point of bifurcation is approached, the difference between the exact and the approximate solutions is so small that it takes a time of order one before the instability mechanism can amplify it to order ϵ. This, however, is not the mechanism for the delay in the Hopf bifurcation. Both these delays can be averted by the presence in the system of noise, even of very low level. For this reason they are probably not of physical significance.

UNFOLDINGS

Consider the simple first-order equation

$$\epsilon \frac{dx}{d\lambda} = (x - \lambda)(x + \lambda) + \delta \equiv G(x, \lambda, \delta) \tag{6}$$

whose equilibrium path is found by setting G equal to zero. If δ vanishes, the solutions are the transcritical curves $x = \pm\lambda$, like those of FIGURE 2(a). If δ is small

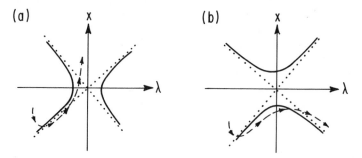

FIGURE 5. Unfolding of the bifurcation of equilibrium curves by the parameter δ. The *dotted curves* are for $\delta = 0$, with bifurcation at the origin. **(a)** $\delta > 0$: the unfolding is such that solutions may escape. **(b)** $\delta < 0$: the unfolding is such that Theorem 1 holds throughout.

but not zero, these curves get separated into a pair of nonintersecting curves, as shown in FIGURE 5(a) and (b). The presence of the additional parameter δ is said to "unfold" the singular pair of curves $x = \pm\lambda$ (singular in that they do not persist under a slight change of the function G) into a pair that is not singular in this sense. Since most physical systems contain features omitted from simple models, it can be argued that the models ought to be robust under small modeling changes.[7] This has led to a conviction, at least among mathematicians, that singular curves should be unfolded in order to achieve realistic behavior. Of course, as FIGURE 5 shows, *how* a curve gets unfolded can qualitatively influence the dynamics.

The term *unfolding* is normally reserved for the situation just described, wherein a singular solution of some equation is desingularized by the effect of parameters. In a sufficiently general understanding of this term, we may say that the singular solutions (solid curves) of FIGURES 2 and 3 are unfolded (dashed curves) by the effect of the small parameter ϵ in the singularly perturbed system (1). Of course, in such a system of differential equations, not all singular solutions are realized in the limit $\epsilon \to 0$, whereas for the algebraic equation $G(x, \lambda, \delta) = 0$ all singular solutions *are* realized in the limit $\delta \to 0$.

We turn now to dynamical systems that are conservative in the limit $\epsilon \to 0$, where we shall find that an unfolding parameter like δ can arise naturally within the model equations considered, that is, without introducing it to make up for some presumed modeling omissions.

CONSERVATIVE DYNAMICS

Consider now the Hamiltonian system

$$\frac{dq}{d\tau} = \frac{\partial H}{\partial p}, \qquad \frac{dp}{d\tau} = -\frac{\partial H}{\partial q}, \tag{7}$$

where the Hamiltonian $H = H(q, p, \lambda)$ depends on the parameter λ, and $\lambda = \epsilon\tau$. This has the form of Eqs. (2) with the vector function $G = (\partial_p H, -\partial_q H)$ and the scalar $F = 1$. If we try to apply Theorem 1, we find we cannot: the eigenvalue condition on the Jacobian matrix G_x fails. This failure is due to the familiar circumstance that a stable equilibrium of a Hamiltonian system cannot be asymptotically stable. Rather a perturbation of the equilibrium position results in oscillations of undiminished amplitude about that position. The full system (7) or (2) need not be conservative for this to be so (since the Hamiltonian in equation (7) is time-dependent, the energy of the system is not conserved); it is only the "frozen" system Eqs. (4), that is conservative. This is sufficient to change the mathematical character of the problem rather completely.

An example is the stellar-evolution problem in an early epoch, before nuclear reactions take place. The evolution is driven by the radiation of energy on the timescale T_L. Passage to the "frozen" system means setting $T_L = \infty$, which amounts to turning off the radiation. With no means of changing its energy, this system is conservative.

We nevertheless anticipate that an analog of Theorem 1 should hold: the solution of the full problem (1) or (2) should perform a kind of oscillatory motion about a stable solution of the reduced problem (3). Here we are interested not in the phase of the oscillation, but in its amplitude. This has a simple and satisfying expression for a Hamiltonian system with one degree of freedom (q and p scalars). Let ϕ, I be angle-action variables, which can always be introduced in the neighborhood of a stable equilibrium point, and let $\omega(I, \lambda) = (\partial H/\partial I)$ be the frequency for the frozen system. Then we have the theorem of adiabatic invariance.[8]

THEOREM 2: Let $\omega(I, \lambda) \neq 0$ in the one-degree-of-freedom system (7). Let $q(\tau, \epsilon)$, $p(\tau, \epsilon)$ be the solution with $\lambda \doteq \lambda_0 + \epsilon\tau$ and initial data q_0, p_0. Then

$$|I(q, p, \lambda) - I(q_0, p_0, \lambda_0)| < C\epsilon \quad \text{for} \quad 0 \leq \tau \leq 1/\epsilon,$$

where C is a positive constant.

This is an entirely satisfactory analog of Theorem 1, but it is restricted to one degree of freedom, whereas Theorem 1 applies in arbitrarily many dimensions.

As soon as we pass to systems with two or more degrees of freedom, the situation deteriorates significantly. The most complete results are of the following form (we refer again to reference 8). Under the assumption that the Hessian matrix $(\partial^2 H/\partial I_i \partial I_k)$ is nonsingular, the change of the action on an interval of length $1/\epsilon$ is proportional, not to ϵ, but to $\sqrt{\epsilon}/\rho$, where ρ is the fraction of the initial phase-space volume excluded from consideration; that is, no estimate is available for this excluded set of initial data. This is an apparently unavoidable consequence of the effects of resonances. It is evident that making ρ small worsens the estimate for the remaining fraction of initial data. Moreover, there is the implicit, and by no means trivial, assumption that the frozen system is completely integrable, so that we can introduce angle-action variables. Nevertheless, this result provides a basis for understanding the behavior of the system near a family of stable equilibriums. At a bifurcation point, these results of adiabatic invariance, whether for one or many degrees of freedom, require modification.

Returning to the case of one degree of freedom, consider the situation depicted in FIGURE 6: There are two equilibriums of system (7) for each value of the

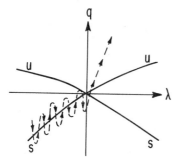

FIGURE 6. For the transcritical bifurcation at $\lambda = 0$, the oscillations carry the phase point into the region of unbounded orbits and, with high probability, the phase point escapes.

parameter λ, one stable and one unstable, that is, this is the transcritical case considered in FIGURE 2. Now, however, the dynamics are different and this is reflected in the oscillatory behavior of a solution of the full system that starts out near an equilibrium point. According to Theorem 2, that solution oscillates about the equilibrium solutions as long as the latter remain safely stable. What happens when the variable λ approaches and then surpasses the bifurcation point can be inferred with some confidence from numerical calculations and related theoretical work.[9-11] The solution deviates very far from either equilibrium branch in a short time, and its description requires full use of the dynamical equations. We emphasize that there is as yet no theory for this on the sound basis characterizing Theorems 1 and 2, and other theorems to which we have made reference. The ideas on which the conclusion is based are indicated in FIGURE 7. If the area enclosed by the orbit (solid curve) remains fixed, that is, if the action is unchanged, the orbit must eventually penetrate the separatrix and become unbounded.

A similar but simpler situation occurs in the case of a pitchfork bifurcation (FIG.

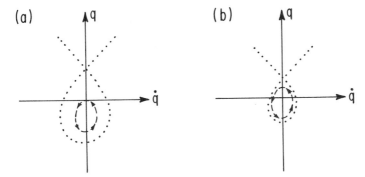

FIGURE 7. The phase portraits for two values of λ when the equilibriums are as in FIGURE 6. The *dotted curves* are separatrices: periodic orbits inside, unbounded orbits outside. **(a)** λ = $\lambda_0 < 0$, a periodic orbit. **(b)** $\lambda_0 < \lambda < 0$; the separatrix has shrunk while the area of the orbit has remained the same. Further shrinkage must result in the phase portrait encountering the region of unbounded orbits.

8). Here the orbit is expected to remain bounded since all orbits of the frozen system are bounded. For values λ surpassing the bifurcation value λ = 0, the orbit may be about one or the other of the two stable equilibriums, or encircle all three. Again, there is no precise theorem covering this situation, but it is close in spirit to the work described in references 9–11.

From the standpoint of fluid dynamical problems, the restriction to one degree of freedom is hopelessly severe. These problems are in principle infinite-dimensional. Even if they can, by various approximations, be reduced to finitely many dimensions, we cannot expect a reduction to one degree of freedom. In particular, if rotation is large enough that Coriolis forces play a role, the minimum number of degrees of freedom that makes sense is two. Nevertheless, the fluid-dynamical problems of astrophysics and of geophysics have, at least in their more idealized forms, a feature that may simplify their behavior and the analysis of their behavior. That feature is the existence of exact integrals of the motion, such as angular momentum and, in idealized examples widely used[12,13] in theoretical astrophysics, circulation integrals conserved by virtue of Kelvin's theorem.

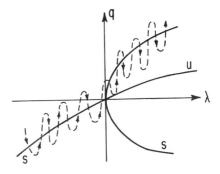

FIGURE 8. For the supercritical pitchfork, the solution may follow one or the other branch, depending on initial data (or may encircle both).

For this reason we consider some "toy" problems in two degrees of freedom possessing a conserved quantity. For the system of equations

$$\frac{d^2x}{dt^2} - 2\omega\frac{dy}{dt} + x(x - \lambda) = 0,$$

$$\frac{d^2y}{dt^2} + 2\omega\frac{dx}{dt} = 0,$$

$$\frac{d\lambda}{dt} = \epsilon, \tag{8}$$

the frozen system ($\epsilon = 0$) has the transcritical equilibrium structure shown in FIGURE 9. The stability of these equilibriums is altered from the previous transcritical cases shown: it does not change at the bifurcation point. The branch stable to the left continues to be (linearly) stable until $\lambda = 4\omega^2$, and the branch stable to the right is stable from $\lambda = -4\omega^2$. This reflects the difference between one and two degrees of

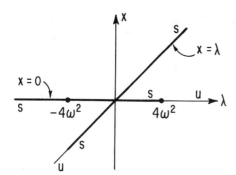

FIGURE 9. Equilibrium solutions of Eq. (8) for $\epsilon = 0$ and different choices of λ. The *darker lines* represent linearly stable branches: for $-4\omega^2 < \lambda < +4\omega^2$, both branches are stable.

freedom, but does not exploit the presence of the exact integral of (8), the quantity $dy/dt + 2\omega x = c$(say). The presence of this integral converts the frozen system corresponding to (8) to an integrable system, whose reduced form is

$$\frac{d^2x}{dt^2} + \frac{dU}{dx} = 0, \tag{9}$$

where

$$U = \tfrac{1}{3}x^3 + (2\omega^2 - \tfrac{1}{2}\lambda)x^2 - 2\omega cx = U(x, \lambda, c). \tag{10}$$

The equivalent one-degree-of-freedom system (9) now contains a further parameter, c, reflecting the original two-degrees-of-freedom system through its conserved quantity. The equilibrium structure of this system depends on the value of c, that is, it depends on the initial data assigned in the original two-degrees-of-freedom system. The families of equilibrium solutions of (9) are shown in FIGURE 10 for $c = 0$ and for two representative values of c of opposite sign. The effect of this parameter is to

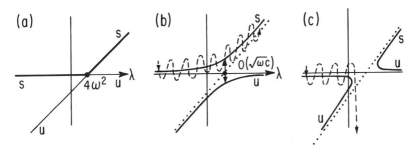

FIGURE 10. Equilibrium solutions of Eq. (9). (a) $c = 0$; *darker lines* represent stable branches. (b) $\omega c > 0$; the bifurcation is unfolded, the upper branch is stable, and Theorem 2 applies if the initial data are small enough. (c) $\omega c < 0$; the phase point must reach the region of unbounded motion.

unfold the bifurcation. Theorem 2 can be used to infer that the motion will take place near the stable equilibriums, provided that the period does not become infinite anywhere along it. It thus appears that in case $\omega c > 0$ [FIG. 10(b)], the motion may follow the analog of the stable branch of equilibrium figures indicated in FIGURE 10(a). These in turn are qualitatively, though not quantitatively, similar to the stable equilibriums of the frozen system corresponding to (7); the latter are what one obtains by seeking stable equilibriums irrespective of conserved quantities. If $\omega c < 0$, dynamical behavior must take place when the parameter λ exceeds zero by some small amount (less than $4\omega^2$). If initial data are essentially unknowable, one might conclude that the two possibilities indicated in FIGURE 10(b) and (c) are equally probable, so that either will occur with probability one-half. This is so provided the initial data are sufficiently small: the gap between stable and unstable behavior is proportional to $\sqrt{|\omega c|}$ so, if c is sufficiently small, that is, if the initial data are sufficiently small, the oscillatory orbit will not bridge the gap.

If in Eq. (7) we replace $x - \lambda$ by $x^2 - \lambda$, the transcritical case of FIGURES 9 and 10 is replaced be the supercritical pitchfork of FIGURES 11 and 12. These indicate the possibilities graphically. The situation is essentially that previously described, except that stable behavior is expected in all cases. For sufficiently small initial data, the trajectory remains close to one or the other stable branch. For larger initial data, it presumably winds about all three equilibriums.

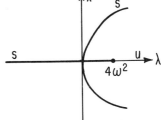

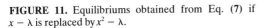

FIGURE 11. Equilibriums obtained from Eq. (7) if $x - \lambda$ is replaced by $x^2 - \lambda$.

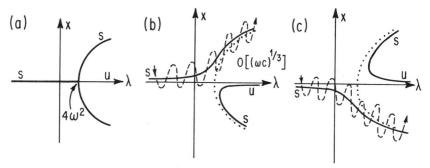

FIGURE 12. Unfolding of the bifurcation diagram of FIGURE 11 by the parameter c. **(a)** $c = 0$. **(b)** $\omega c > 0$. **(c)** $\omega c < 0$.

SELF-GRAVITATING FLUIDS

The application of ideas from ordinary differential equations to problems in fluid dynamics is possible if the partial differential equations of the latter subject can be reduced to a system of ordinary differential equations. This is a goal that is being widely pursued at present in a variety of fluid-dynamical contexts. One context in which the goal is achieved rather simply is that of the motion of self-gravitating fluids of uniform density preserving an ellipsoidal shape. The unfolding of bifurcations because of the presence of constants of the motion that serve as unfolding parameters is well illustrated by this example.

In the special class of motions for which the x_3-axis of the ellipsoid remains aligned with the angular velocity vector, the velocity components may be written[14]

$$u_1 = \frac{\dot{a}_1}{a_1} x_1 + \mu \frac{a_1}{a_2} x_2, \qquad u_2 = \frac{\dot{a}_2}{a_2} x_2 - \mu \frac{a_2}{a_1} x_1, \qquad u_3 = \frac{\dot{a}_3}{a_3} x_3, \qquad (11)$$

where a_1, a_2, and a_3 are the semiaxes, μ is a variable representing the vorticity of the figure relative to the rotating frame, and the overdots indicate time differentiation. The dynamical equations are derivable from a Lagrangian

$$L = \tfrac{1}{2} \left(\dot{a}_1^2 + \dot{a}_2^2 + \dot{a}_3^2 \right) + \tfrac{1}{2} \left(a_1^2 + a_2^2 \right) \left(\omega^2 + \mu^2 \right) - 2\omega\mu a_1 a_2 + I(a_1, a_2, a_3), \qquad (12)$$

where I is a function representing the gravitational potential of an ellipsoid of uniform density. Two coordinates, say θ and ϕ where $\omega = \dot{\theta}$ and $\mu = \dot{\phi}$, are ignorable, so the following quantities are constant:

$$\frac{\partial L}{\partial \omega} = \left(a_1^2 + a_2^2 \right) \omega - 2 a_1 a_2 \mu,$$

$$\frac{\partial L}{\partial \mu} = \left(a_1^2 + a_2^2 \right) \mu - 2 a_1 a_2 \omega, \qquad (13)$$

FIGURE 13. Steady-state Riemann ellipsoids with velocity field given by Eq. (11) are found in the region between the two solid curves. The Jacobi family is indicated by the *dashes*. Along the Maclaurin family ($a_2/a_1 = 1$) and along the lower boundary, $K_1 = 0$.

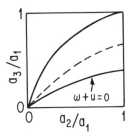

or

$$(a_1 - a_2)^2 (\omega + \mu) = K_1 \quad \text{and} \quad (a_1 + a_2)^2 (\omega - \mu) = K_2. \tag{14}$$

The possible steady-state figures of these Riemann ellipsoids are indicated in FIGURE 13, which shows the region that they occupy in the plane of a_2/a_1 versus a_3/a_1. The dotted line shows the Jacobi ellipsoids, which are the figures one would obtain by setting all time derivatives equal to zero in the equations of motion, rather than in the reduced equations obtained by first isolating the constants of the motion. The Jacobi family is therefore analogous to the (somewhat misleading) family of FIGURE 9. The lower boundary of the occupied region corresponds to the value $K_1 = 0$. It represents the so-called self-adjoint ellipsoids, obtained if one *does* take the constants of motion into account. FIGURE 14 shows the bifurcation diagram and its unfolding by the parameter K_1 in the plane of $(a_1 - a_2)/(a_1 + a_2)$ versus μ. It reveals the same features previously seen in the artificial problem (8) and indicated in FIGURE 10: the unfolding of the bifurcation implies that the dynamical behavior may occur smoothly, without any "catastrophic" changes, even in the vicinity of the bifurcation point. Whether it follows the stable branch of equilibriums of the reduced problem or not depends on the size of the initial data. If the latter are large, so that the the excursions in phase space are larger than the gaps between branches, a sudden change may indeed occur. The details of this transition are not understood even in the simplest cases.

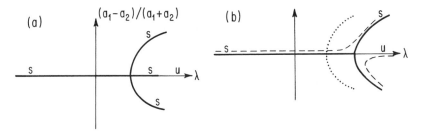

FIGURE 14. (a) Obtained from the steady-state equations without accounting for constants of the motion. The Jacobi family bifurcates from the Maclaurin family. (b) Obtained from the reduced problem accounting for constants of the motion. The Jacobi family is *dotted*. The lower self-adjoint family bifurcates for $K_1 = 0$. For $K_1 \neq 0$ the bifurcation is unfolded (*dashed curves*).

These results bear on the formulation of the fission theory of binary-star formation. The traditional formulation of this problem, which goes back to Thomson and Tait,[15] effectively assumes a smooth transition from axial symmetry to the lower symmetry expressed by ellipsoids with three unequal axes, followed later in the evolution by a further transition representing the initial stages of an incipient binary star. This formulation has been abandoned by some of those investigating the possibility of fission numerically[12,13] in favor of the following hypothesis[16]: The bifurcation signals a catastrophic event leading directly to the production of a binary star. The numerical calculations based on this hypothesis cannot be done directly at the bifurcation point for reasons of cost and accuracy, and are therefore initiated somewhere beyond (in FIG. 14, to the right of) the bifurcation point. Binary formation is not observed in these simulations, and the conclusion is drawn that the fission mechanism does not work. But this conclusion is flawed because the premise is flawed: there is no reason to expect *any* catastrophic event at the bifurcation point. Moreover, the transition taking place near that point would appear to be in the direction of triaxial figures, and this is consistent with the numerical results. The latter can therefore be interpreted as favoring the traditional formulation, with no implication one way or the other on whether the ultimate outcome of the evolution is fission.

REFERENCES

1. TIKHONOV, A. N. 1952. Mat. NS(31) **73**: 575. (In Russian.)
2. LEVIN, J. & N. LEVINSON. 1954. J. Rat. Mech. Anal. **3**: 247.
3. LEBOVITZ, N. & R. SCHAAR. 1975. Stud. Appl. Math. **LIV**: 229.
4. ———. 1977. Stud. Appl. Math. **LVI**: 1.
5. NEISHTADT, A. I. 1987. Differ. Equations **23**: 1385.
6. BAER, S. M., T. ERNEUX & J. RINZEL. 1989. SIAM J. Appl. Math. **49**(1): 55.
7. GOLUBITSKY, M. & D. SCHAEFFER. 1984. Singularities and Groups in Bifurcation Theory. Springer-Verlag. New York/Berlin.
8. ARNOL'D, V. I. 1978. Mathematical Methods of Classical Mechanics. Springer-Verlag. New York/Berlin.
9. NEISHTADT, A. I. 1986. Phys. Plasmy **12**: 922.
10. CARY, J., D. F. ESCANDE & J. L. TENNYSON. 1986. Phys. Rev. A. **34**(5): 4256.
11. HANNAY, J. H. 1986. J. Phys. A: Math. Gen. **19**: L1067.
12. DURISEN, R., J. TOHLINE & M. MCCOLLOUGH. 1985. Astrophys. J. **298**: 220.
13. ERIGUCHI, Y., I. HACHISU & D. SUGIMOTO. 1982. Prog. Theor. Phys. **67**: 1068.
14. CHANDRASEKHAR, S. 1987. Ellipsoidal Figures of Equilibrium. Dover. New York.
15. THOMSON, W. & P. G. TAIT. 1883. Treatise on Natural Philosophy, Vol. I, Part 2. Cambridge University Press. Cambridge, England.
16. OSTRIKER, J. P. 1970. *In* Stellar Rotation, A. Slettebak, Ed.: 147. Gordon & Breach. New York.

New Results on Discrete Map Simulations of Nonlinear Pulsations[a]

J. PERDANG

Institute of Astronomy
Madingley Road
Cambridge CB3 OHA, England
and
[b]Institut d'Astrophysique
5 Avenue de Cointe
Cointe-Ougrée B4200, Belgium

INTRODUCTION

The models of stellar oscillations so far investigated theoretically are all predicated on the assumption that the stellar oscillator is equivalent to a low-dimensional dynamical system. The motions supported by this oscillator are therefore periodic oscillations, multiperiodic oscillations (with a small number of basic periods), bifurcations of the latter, and deterministic chaos. The hydrocode experiments carried out with realistic models[1,2] indicate that the underlying equations are indeed equivalent to a very low-dimensional dynamical system: from the estimate of the dimension of the attractor of the hydrocode solutions, one infers that a third-order differential system should be sufficient to describe the oscillatory properties of both population I and population II stars in the instability strips. The conclusion is that just a single linear mode enters the actual dynamical behavior (a single dissipative mode, or equivalently, a one-zone model described by a differential equation of third order). A third-order differential system, however, discards the effect of resonances between modes that have been observed to play a decisive part in the bifurcations (cf. Buchler[3]). While a three-dimensional differential dynamical system may be sufficient to reproduce the qualitative time behavior of stellar oscillations as generated by current hydrocodes of *fixed model parameters* (fixed mass, composition, luminosity, etc.), the three-dimensional differential dynamical system cannot be adequate to capture the *transitions* from one type of time behavior to another under *continuous parameter changes*. A higher order differential system is required to this end.

Since it is out of the question to perform a reasonably exhaustive survey of the time behavior of the bounded solutions of parametrized higher order differential systems, we have tried to construct a much simpler dynamical system, namely a discrete parametrized map, which hopefully preserves the main qualitative features of higher order differential systems. Such a discrete map can be derived from the standard stellar hydrodynamic equations (cf. Perdang,[4] hereinafter referred to as Paper I). In the present paper we shall attempt to set up a discrete family of maps in a formal way that does not hinge on the currently adopted physics of stellar hydrody-

[a]This work was supported by a Royal Society-FNRS European Exchange Fellowship (1989).
[b]Where correspondence should be addressed.

87

namics. The new family can then also model, in principle, classes of mechanisms not included in the standard stellar hydrodynamic equations (mass-loss, magnetic effects, etc.). Ideally, the family of maps should possess the property that if a member of the family describes the oscillation of a given real star (or realistic stellar model) then *any* slight change in the physics of the star should again be described by a "nearby" member of the same family of maps. In the language of catastrophe theory, this property requires our maps to be *structurally stable*.[5-7] If we simulate the behavior of a class of stars by a family of maps, then the property of structural stability secures, by definition, that all observable oscillations this class can exhibit in the neighborhood of a given oscillatory state are actually encountered in the model family.

The preceding ambitious concept of a fully structurally stable family of maps simulating the time behavior of a given class of variable stars has not been investigated. As a substitute we have introduced a restricted concept of structural stability with respect to a preassigned class of perturbation functions that, on the one hand, is broad enough to isolate what we think are the most relevant types of oscillations, and on the other remains operationally feasible.

STRUCTURAL STABILITY OF ITERATIVE MAPS $S_n[\mathbf{F}]$ AND FAMILIES OF ITERATIVE MAPS, $S_n[\mathbf{F}[\mathbf{y};\lambda]]$

In Paper I a parametric two-dimensional discrete map was generated from the standard partial differential equations of stellar fluid dynamics. The main assumptions in the derivation are the following: (1) a single mode dominates the oscillation process; and (2) the coupling of this mode with the (infinity of) remaining modes is described by periodic coefficient functions relating the unknown amplitudes of the remaining modes to the amplitude of the mode we are dealing with. Through this procedure we do include the possibility of transitions induced by mode interactions. By explicitly isolating a single mode as being dominant we exclude double-mode oscillations at the outset. The purpose of the present section is to develop a discrete map representative for the stellar oscillations within a broader framework in which we avoid the specific physical assumptions of our original approach. Instead, our present formulation centers around a concept of structural stability of the maps.

We consider an n-dimensional discrete-time map, $S_n[\mathbf{F}]$,

$$\mathbf{y}(t + 1) = \mathbf{F}[\mathbf{y}(t)], \qquad t = 0, 1, 2, \ldots, \tag{1}$$

with

$$\mathbf{y}(t) = (y_1(t), y_2(t), \ldots, y_n(t)), \tag{1a}$$

$$\mathbf{F}[\mathbf{y}(t)] = (F_1[y_1(t), y_2(t), \ldots, y_n(t)],$$
$$F_2[y_1(t), y_2(t), \ldots, y_n(t)], \ldots, F_n[y_1(t), y_2(t), \ldots, y_n(t)]), \tag{1b}$$

defined for $\mathbf{y} \in \mathbb{D}^n$ a (compact) subset of the Euclidian space $\mathbb{E}^n$, and for $F_i[\mathbf{y}] \in C^\infty(\mathbb{D}^n)$, $i = 1, 2, \ldots, n;$ $\mathbf{F} : \mathbb{D}^n \to \mathbb{D}^n$. The set $\mathbb{D}^n$ will be referred to as the behavior space.

Consider the partition of $\mathbb{D}^n, \mathscr{P}\{\mathbb{D}^n\}$:

$$\mathbb{D}^n = \cup_T \mathbb{D}^n(T), \quad \mathbb{D}^n(T) \cap \mathbb{D}^n(L) = \varnothing \quad \text{if} \quad T \neq L, \tag{2}$$

where $\mathbb{D}^n(T)$ is the basin of a given type T of time behavior (an eventually or asymptotically period-K oscillation $(P\text{-}K)$, $K = 1,1', \ldots ; 2,2',2'', \ldots ; 3,3',3'', \ldots ;$ $\ldots$ or a chaotic (C) solution, $c,c',c'', \ldots$, respectively, of the map (1); for any $\mathbf{y}(0) \in \mathbb{D}^n(T)$ we have $\mathbf{y}(t) \in \mathbb{D}^n(T)$ for $t = 1,2,3, \ldots$. Two basins $\mathbb{D}^n(T)$ and $\mathbb{D}^n(L)$ are regarded as distinct if the associated attractors $\mathbb{A}(T)$ and $\mathbb{A}(L)$ are different. For any initial condition $\mathbf{y}(0) \in \mathbb{D}^n(T)$ we have $\mathbf{y}(t) \in \mathbb{D}^n(T)$ for $t = 1,2,3, \ldots$ and $\mathbf{y}(t) \in \mathbb{A}(T) \subset \mathbb{D}^n(T)$ for $t \to \infty$. By eventually $P\text{-}K$ we mean that the solution becomes $P\text{-}K$ after a finite number of steps, that is, the attractor $\mathbb{A}(T)$ is reached in a finite time; for an asymptotically $P\text{-}K$ solution, the $P\text{-}K$ behavior, that is, the attractor $\mathbb{A}(T)$, is approached asymptotically (cf. Perdang,[8] cf. also Lauwerier[9] for the one-dimensional case).

The map $S_n[\mathbf{F}]$ is said to be *structurally stable* if for any C^∞ correction function $\mathbf{f}$ ($\|\mathbf{f}\|$ ϵ, the norm being the C-norm; cf. below; ϵ a preassigned precision) such that

$$\mathbf{F} + \mathbf{f}: \mathbb{D}^n \to \mathbb{D}^n, \qquad \mathbf{F}, \mathbf{f} \in C^\infty, \tag{3}$$

the partition of the bounded region $\mathbb{D}^n, \mathscr{P}\{\mathbb{D}^n\}$, into invariant $P\text{-}K$ and c subsets, $\mathbb{D}^n(T)$, of the corrected map $S_n[\mathbf{F} + \mathbf{f}]$,

$$\mathbb{D}^n = \cup_T \mathbb{D}^n(T), \quad \mathbb{D}^n(T) \cap \mathbb{D}^n(L) = \varnothing \quad \text{if} \quad T \neq L, \tag{4}$$

is such that there exists a homeomorphism H between the two partitions $\mathscr{P}\{\mathbb{D}^n\}$ and $\mathscr{P}\{\mathbb{D}^n\}$

$$H : \mathscr{P}\{\mathbb{D}^n\} \to \mathscr{P}\{\mathbb{D}^n\} . \tag{5}$$

We then say that the original map $S_n[\mathbf{F}]$ and the perturbed maps $S_n[\mathbf{F} + \mathbf{f}]$ have *equivalent time behavior*. If no homeomorphism (5) can be found, then the map $S_n[\mathbf{F}]$ is *structurally unstable*.

The equivalence of time-behavior of two maps involves essentially three elements: (a) the two maps show the same types of oscillations; (b) if a given type T of oscillation has "zero probability" to occur for map $S_n[\mathbf{F}]$ (i.e., the Euclidean measure of the corresponding set $\mathbb{D}^n(T)$ is zero), then it also has zero probability in the perturbed maps $S_n[\mathbf{F} + \mathbf{f}]$; and (c) contiguous domains of different types of solution in $\mathbb{D}^n$ remain contiguous in $\mathbb{D}^n$; if we continuously change the initial condition $\mathbf{y}(0)$ over a curve $\mathscr{C} \subset \mathbb{D}^n$, producing successively solutions of type $A, B, \ldots$, then there exists a corresponding curve $\mathscr{C}' \subset \mathbb{D}^n$ such that if we continuously change the initial conditions along that curve we again successively generate solutions of type $A, B, \ldots$.

To show that a given map $S_n[\mathbf{F}]$ is structurally unstable it suffices to exhibit a single perturbation function $\mathbf{f}$ that leads to the violation of condition (5). For instance, the one-dimensional map

$$y(t + 1) = y(t)^2 + y(t), \tag{6}$$

defined over the interval $\mathbb{D} = [-a, +a]$, a a small positive real, is structurally unstable. Under the linear perturbation $f_1 = \epsilon y$ we produce the map

$$y(t + 1) = (1 + \epsilon) y(t) + y(t)^2, \tag{6a}$$

(the canonical form of the *transcritical bifurcation*, cf. Lauwerier[9]). Under the perturbation by a constant disturbance, $f_2 = -\eta$

$$y(t + 1) = -\eta + y(t) + y(t)^2, \qquad (6b)$$

(the *fold bifurcation*, cf. Lauwerier[9]). The topology of the fixed points of (6), (6a), and (6b) is seen to be different; the partition of the domain of definition $\mathbb{D}$ of the map thus changes under infinitesimal perturbations f_1 or f_2, in violation with (5).

The only constraint imposed on the perturbing function **f** is *smoothness* [Eq. (3)]. This requirement allows us to represent any admissible correction **f** in the neighborhood of any point y_0 of the behavior space in the form of a power series in $(y - y_0)$

$$\mathbf{f}[\mathbf{y}] = \mathbf{a} + B \cdot (\mathbf{y} - \mathbf{y}_0) + C \cdot\cdot (\mathbf{y} - \mathbf{y}_0)(\mathbf{y} - \mathbf{y}_0) + \cdots, \qquad (7)$$

(where **a** is a constant n-vector; B an $n \times n$ matrix, C an $n \times n \times n$ matrix, etc.; the dot $\cdot$ stands for the scalar product), so that the whole class of smooth perturbations defines a family of functions of a (countable) infinity of parameters $(a_i, B_{ij}, C_{ijk}, \ldots, i, j, k, \ldots = 1, 2, \ldots, n)$.

Since we cannot carry out numerically an infinity of tests, we make use of a weaker concept of structural stability that concentrates on a *finite* subset of perturbation functions, hopefully to be chosen general enough to capture the most relevant physics. To characterize this restricted class of admissible perturbation functions **f** we propose the following procedure. The original map $S[\mathbf{F}]$ we analyze is required to represent already a first-order approximation to the true stellar pulsations. By this we mean that the possible differences **f** between the approximate model and the real star are not only small in amplitude, $\|\mathbf{f}\| < \epsilon$, where ϵ is some preassigned tolerance, the norm being chosen as the C-norm

$$\|\mathbf{f}\| \equiv \|\mathbf{f}\|_0 = \max_{\mathbf{y} \in \mathbb{D}} |\mathbf{f}[\mathbf{y}]|, \qquad (8)$$

where $|A|$ denotes the modulus of the largest element of A (a vector, matrix, etc.); the differences **f** do not fluctuate too wildly with **y** either. The latter property means that the norm to be chosen is a D_m-norm, m some positive integer,

$$\|\mathbf{f}\|_m \equiv \|\mathbf{f}\|_0 + \|\partial/\partial\mathbf{y}\,\mathbf{f}\|_0 + \cdots + \|\partial^m/\partial\mathbf{y}\partial\mathbf{y}\cdots d\mathbf{y}\,\mathbf{f}\|_0. \qquad (9)$$

In fact, with respect to sufficiently strongly fluctuating perturbing functions a map is always structurally unstable, as is made clear by a one-dimensional example: Choose $S_1[F]$ to have an isolated fixed point at $y^* = 0$; then the perturbation $f = \epsilon \sin(Ny)$ generates a large number of fixed points close to y^*, provided only that N is large enough. If such a perturbation were physically acceptable, it would only imply that the reference model itself was an inadequate approximation to the realistic problem it was meant to simulate, in contradiction with our initial assumption. The C-norm is here ϵ, while the D_m-norm is of the order of ϵN^m, which, for a fixed ϵ, can be made as large as we wish by taking N large enough. According to standard knowledge, in the neighborhood of any realistic stellar equilibrium we never have a large number of other equilibrium states (or distinct oscillation states of the same linear period);

therefore, we must require that the admissible maps have a small number of fixed points. This requirement is fulfilled if the perturbation functions are chosen to be polynomials of a low degree d

$$\mathbf{f}[\mathbf{y}] \equiv \mathbf{P}_d[\mathbf{y}] = \mathbf{a} + B \cdot (\mathbf{y} - \mathbf{y}_0) + C \cdot\cdot (\mathbf{y} - \mathbf{y}_0) (\mathbf{y} - \mathbf{y}_0)$$

$$+ \cdots + D \cdot\cdot\cdot\cdot\cdot (\mathbf{y} - \mathbf{y}_0) (\mathbf{y} - \mathbf{y}_0) \cdots (\mathbf{y} - \mathbf{y}_0), \quad (10)$$

where the last term contains d products of $(\mathbf{y} - \mathbf{y}_0)$. The most general perturbation function involves a finite number $q(d, n)$ of parameters, $q(d, 1) = d + 1$; $q(d, 2) = (d + 1)(d + 2); \ldots$. The space of functions (10), P_d^q, is a q-dimensional vector space spanned by $\mathbf{e}_1 = (1, 0, 0, \ldots, 0)$, $\mathbf{e}_2 = (0, 1, 0, \ldots, 0), \ldots, \mathbf{e}_n = (0, 0, \ldots, 1)$, $\mathbf{e}_{n+1} = (y_1 - y_{10})\mathbf{e}_1$, $\mathbf{e}_{n+2} = (y_1 - y_{10})\mathbf{e}_2, \ldots, \mathbf{e}_{n+n} = (y_1 - y_{10})\mathbf{e}_n, \ldots, \mathbf{e}_q = (y_n - y_{n0})^d \mathbf{e}_n$.

With these preliminaries we now define the restricted concept of d-*structural instability* of an n-dimensional map $S_n[\mathbf{F}]$: The map $S_n[\mathbf{F}]$ specified by Eqs. (1) and (2) is said to be d-*structurally stable* if for any small correction function $\mathbf{f} \in P_d^q$ such that

$$\mathbf{F} + \mathbf{f} : \mathbb{D}^{\cdot n} \to \mathbb{D}^{\cdot n}, \quad (11)$$

the partition of the bounded compact set $\mathbb{D}^{\cdot n}, \mathscr{P}\{\mathbb{D}^{\cdot n}\}$, into invariant basins, $\mathbb{D}^{\cdot n}(T)$, $(T = 1, 1', \ldots, 2, 2', \ldots, 3, 3', \ldots$, and $c, c' \ldots)$, of the corrected map $S_n[\mathbf{F} + \mathbf{f}]$, is such that there exists a homeomorphism H, Eq. (5), between the two partitions $\mathscr{P}\{\mathbb{D}^n\}$ and $\mathscr{P}\{\mathbb{D}^{\cdot n}\}$. The two maps $S_n[\mathbf{F}]$ and $S_n[\mathbf{F} + \mathbf{f}]$ are then said to have d-*equivalent time behavior*. Otherwise, the map $S_n[\mathbf{F}]$ is d-*structurally unstable*.

For d small enough, the number of parameters involved in the polynomial perturbing functions is small, so that we can explicitly test for d-structural stability by numerical experiments.

We have so far considered individual maps $S_n[\mathbf{F}]$ only. For the purposes of describing variable stars we are actually interested in maps depending on *parameters*, $\lambda = (\lambda_1, \lambda_2, \ldots, \lambda_N) \subset \Lambda^N \subset \mathbb{E}^N$, where N is finite, to be denoted by $S_n[\mathbf{F}[\mathbf{y};\lambda]]$. The parameters simulate either the effect of changes occurring in an individual star in the course of its evolution, or else the physical differences between individual stars of a class (differences in mass, composition, etc.), or possibly between classes of variable stars (say ZZ Ceti and δ Scuti stars, high-luminosity and low-luminosity long-period variables, etc.). The construction of these parametrized maps will be guided by the concept of d-*structural stability* of a family $S_n[\mathbf{F}[\mathbf{y};\lambda]]$ defined as follows: Consider an N-parameter family of n-dimensional maps $S_n[\mathbf{F}[\mathbf{y};\lambda]]$ defined over the control space $\mathbb{C}^{n+N} \subset \mathbb{E}^{n+N}$ (domain of definition of the behavior variables, $\mathbf{y} \in \mathbb{D}^n$ in general depending on λ, and of the parameters, $\lambda \in \Lambda^N$); each individual map of this family is specified by Eqs. (1) and (2). Let $\mathscr{P}\{\mathbb{C}^{n+N}\}$ be the partition of the control space

$$\mathbb{C}^{n+N} = \cup_T \mathbb{C}^{n+N}(T), \qquad \mathbb{C}^{n+N}(K) \cap \mathbb{C}^{n+N}(L) = \varnothing \quad \text{if} \quad K \neq L \quad (12)$$

into basins $\mathbb{C}^{n+N}(T)$ of solutions of type T (period-K oscillations, chaotic oscillations). The family is said to be d-*structurally stable* if for any small enough correction function $\mathbf{f} \in P_d^q$ such that

$$\mathbf{F} + \mathbf{f} : \mathbb{C}^{rn+N} \to \mathbb{C}^{rn+N}, \tag{13}$$

the partition of the bounded compact set $\mathbb{C}^{rn+N}$, $\mathscr{P}\left\{\mathbb{C}^{rn+N}\right\}$, into invariant basins of type T, $\mathbb{C}^{rn+N}(T)$, $(T = 1, 1', \ldots, 2, 2', \ldots, 3, 3', \ldots,$ and c, c', $\ldots)$, of the perturbed family $S_n[\mathbf{F}[\mathbf{y};\lambda] + \mathbf{f}]$, is such that there exists a homeomorphism H^* between $\mathscr{P}\left\{\mathbb{C}^{n+N}\right\}$ and $\mathscr{P}\left\{\mathbb{C}^{rn+N}\right\}$

$$H^* : \mathscr{P}\left\{\mathbb{C}^{n+N}\right\} \to \mathscr{P}\left\{\mathbb{C}^{rn+N}\right\}. \tag{14}$$

Otherwise, the family of maps $S_n[\mathbf{F}[\mathbf{y};\lambda]]$ is d-*structurally unstable*.

Since it is among the structurally unstable maps that transitions in the time behavior occur [cf. the one-dimensional example Eq. (6)], a family of maps designed to mimic a broad range of stellar pulsational behavior must include structurally unstable maps. The family itself is naturally required to be d-structurally stable: By definition, for some admissible perturbations, an unstable family would display classes of time behavior not accounted for in the framework of the family; the star of which the family of maps is thought to be representative would then produce new patterns of behavior not simulated by the family, under some infinitesimal changes in its physical properties (say, slight changes in the chemical composition, etc.). In contrast, the d-structurally stable family of maps, even though only a substitute for detailed hydrodynamic models, reveals, ideally, all allowed types of time behavior of the realistic star.

Starting out with a d-structurally unstable individual map $S_n[\mathbf{F}_u]$, by including a large enough number N of perturbations $\mathbf{f}_1, \mathbf{f}_2, \ldots, \mathbf{f}_N$, we generate a d-structurally stable N-parameter family of maps

$$\mathbf{F}[\mathbf{y}; \lambda] = \mathbf{F}_u[\mathbf{y}] + \lambda_1 \mathbf{f}_1[\mathbf{y}] + \lambda_2 \mathbf{f}_2[\mathbf{y}] + \cdots + \lambda_N \mathbf{f}_N[\mathbf{y}]. \tag{15}$$

Note that the family (15) is defined and remains stable for *finite* parameters λ; therefore, we are entitled to expect that it is representative not only for stars differing only slightly from the structure of the original structurally unstable variable star, but for a whole class of variable stars. We shall adopt procedure (15) for stabilizing an unstable map in our construction of families of maps representative for stellar pulsations. Such an approach involves two basic unknowns: (1) What is the degree d of the polynomials with respect to which we should require structural stability? In our applications we have set $d = 2$, allowing for the lowest order nonlinear perturbations only. This choice secures that the number of equilibrium models close to the original model remains indeed small. In the remainder, the designation *structural stability* will stand as an abbreviation for *2-structural stability*. We leave it to later investigations to inquire into the role of higher order structural stability. (2) How do we choose the original unstable model? We select the algebraically simplest maps of given dimension. Whether choices (1) and (2) are acceptable for whole classes of variables can only be decided *a posteriori*.

GENERATING ONE- AND TWO-DIMENSIONAL FAMILIES OF MAPS OF STELLAR PULSATIONS

One-dimensional Family

The algebraically simplest iteration generating an oscillation (P-1 for any initial condition $y(0) = y_0 \in \mathbb{E}^1$) is the trivial map $S_1[y]$

$$y(t + 1) = y(t), \tag{16}$$

We can view this map as simulating a stationary linear stellar oscillation, where y stands for the amplitude of a stellar observable (radial velocity, radius, etc.). This map is structurally unstable, since under the perturbation $f[y] = \epsilon y$, the P-1 oscillation of arbitrary amplitude decays into a P-1 solution of zero amplitude (the star ceases oscillating) if $\epsilon < 0$; for $\epsilon > 0$ the amplitude grows without bounds (and the map ceases to represent a physically realistic oscillation). Perturb the unstable map (16) by an arbitary element of the function space P_2^3 (spanned by 1, y, y^2), to obtain the family $S_1[F[y;\lambda]]$

$$F[y; \lambda] = y + \lambda_1 + \lambda_2 y + \lambda_3 y^2. \tag{17}$$

This family is (2-) *structurally stable* for *any* values of the parameters, not just for arbitrarily small values, since any perturbation of (17), for any λ, by an element of the space P_2^3 generates the same family (the parameters being redefined). Among the three arbitrary parameters two parameters are irrelevant, since the transformation $y = Az + B$ applied to the iterations $y(t + 1) = y(t) + \lambda_1 + \lambda_2 y(t) + \lambda_3 y(t)^2$ produces a 1-parameter map, the *transcritical bifurcation*

$$z(t + 1) = (1 + p) z(t) + z(t)^2, \tag{18}$$

if we properly select the free coefficients A and B; p is an essential parameter, in the sense that it cannot be eliminated without losing some time patterns of our family (17). Through a further change of scale (18) is transformed into the *logistic map*

$$Z(t + 1) = (1 + p) Z(t)(1 - Z(t)). \tag{19}$$

Our construction thus shows that the structurally stable one-dimensional family of maps is a 1-parameter family, materialized by the logistic map (19). The one-dimensional model then predicts that the oscillations of genuine stars, or the hydrocode oscillations of realistic stellar models, should show, under certain circumstances, the well-known behavior of the logistic map. Under an increase of the unique parameter p from 0 to 3 the system goes through the standard sequence of Feigenbaum period doublings followed by chaos ($p > p_{chaos} = 2.5699 \ldots$). The irreducible parameter p is essentially the linear growth-rate κ times the linear period P of the oscillation (more precisely $p = \exp(\kappa P) - 1$, cf. Paper I). The physical circumstances under which this one-dimensional map is a fair approximation to the stellar pulsations are reasonably clear: The global oscillation of the star can be approximated by a single mode [zone] that does not interact with the remaining modes [zones]; the time behavior is then described by a system of three ordinary differential equations. The equations of realistic stellar hydrodynamics do not *globally* admit of such a reduction (for obvious group theoretical reasons; cf. Olver[10]). The estimates of the dimensions of the attractors as obtained in the hydrocode experiments by Kovacs and Buchler[2] indicate only that *locally* a reduction to a third-order differential system may hold. Under changes of the physical parameters of the star, due, for instance, to stellar evolution, interactions between different modes are bound to occur (in particular through resonances); the treatment of the latter requires higher dimensional behavior spaces.

Two-dimensional Family

We recall that two-dimensional area-preserving maps are mathematically equivalent to Hamiltonian systems of 2 degrees of freedom (adiabatic 2-mode interactions). By relaxing the requirement of area preservation, the discrete two-dimensional map is made to incorporate dissipation. It should therefore simulate nonadiabatic two-mode interactions. The algebraically simplest two-dimensional map is the trivial map $S_2[\mathbf{y}]$

$$y_i(t + 1) = y_i(t), \qquad i = 1, 2, \tag{20}$$

the direct product of two one-dimensional maps of type (16), $S_2[\mathbf{y}] = S_1[y_1] \times S_1[y_2]$. The new map describes two independent P-1 oscillations for any initial condition $y_i(0) = y_{i0} \in \mathbb{E}^1$, $i = 1, 2$, which we can visualize as the linear oscillations of two uncoupled stellar zones. The direct product of structurally unstable maps is structurally unstable.

Perturb the unstable map (20) by an arbitrary element of the function space P_2^{12} of basis $\mathbf{e}_i, y_1 \mathbf{e}_i, y_2 \mathbf{e}_i, y_1^2 \mathbf{e}_i, y_1 y_2 \mathbf{e}_i, y_2^2 \mathbf{e}_i, i = 1, 2$, to generate a family $S_2[\mathbf{F}[\mathbf{y}; \lambda]]$

$$F_1[\mathbf{y}; \lambda] = y_1 + \lambda_1 + \lambda_2 y_1 + \lambda_3 y_1 + \lambda_4 y_1^2 + \lambda_5 y_1 y_2 + \lambda_6 y_2^2,$$

$$F_2[\mathbf{y}; \lambda] = y_2 + \lambda_7 + \lambda_8 y_1 + \lambda_9 y_1 + \lambda_{10} y_1^2 + \lambda_{11} y_1 y_2 + \lambda_{12} y_2^2. \tag{21}$$

The latter is (2-) *structurally stable* by construction for any $\lambda \in \mathbb{E}^{12}$. By a translation in the behavior space we eliminate two parameters (λ_1, λ_7). A linear homogeneous transformation enables us to diagonalize the coefficient matrix of the linear terms, thereby again eliminating two parameters; here two basically distinct cases arise, depending on whether the two eigenvalues, m_1, m_2, of the matrix are real or complex (conjugate). Finally, by rescaling the behavior variables we can eliminate two further parameters. A nonlinear transformation could eliminate additional parameters, if we limited ourselves to the behavior around $\mathbf{y} = 0$ with fixed eigenvalues m_1, m_2 (cf. the procedure used in the theory of normal forms[11]). Since we are interested in a family of maps defined over a range of the eigenvalues, a nonlinear reduction of the parameters does not seem to be acceptable. Accordingly, we are led to keep six basic parameters $\mathbf{p} = (p_1, p_2, p_3, p_4, p_5, p_6)$, chosen to define the following two-dimensional family, $S_2(p_1, p_2, p_3, p_4, p_5, p_6)$

$$z_1(t + 1) = (1 + p_1)z_1 + z_1^2 + p_2 z_1 z_2 + p_3 z_2^2,$$

$$z_2(t + 1) = (1 + p_4)z_2 + z_2^2 + p_5 z_1 z_2 + p_6 z_2^2. \tag{22}$$

which is adequate if the eigenvalues are *real*. Or we may introduce a complex behavior variable, $q = z_1 + iz_2$, and write the family in the form

$$q(t + 1) = (1 + a)q(t) + q(t)^2 + bq(t)^{*2} + cq(t) q(t)^*, \qquad (23)$$

a, b, c being, in turn, complex parameters ($a = a_R + ia_I$, etc.); this form is adequate if the eigenvalues (m_1, m_2) are *complex*.

Representation (23) is almost identical with the family directly derived from the stellar fluid dynamics (Paper I, eq. 2.24), except that the parameter a in Eq. (23) is complex, while its counterpart in Paper I is a real parameter. The family of maps derived in Paper I thus involves only five essential parameters instead of six. We have therefore reason to believe that it must be (2-) structurally unstable, although a formal proof of this point is lacking. Physically, the requirement in Paper I that a single mode dominate the time behavior excludes resonances, and reduces the residual modes (mechanical or thermal) to play a subsidiary part, even though it is strong enough to induce chaos. In the new algebraic formulation the two modes [or zones] are treated strictly on the same footing; the 6-parameter family can therefore fully deal with a strong excitation of two modes, as well as with resonances among the latter. A minor modification of the family of Paper I thus produces a new family that we believe covers a broader class of stellar oscillations and transitions. Notice that the new family can be generated directly from the family of Paper I by applying the perturbation scheme (15) to the latter.

The family (23) only takes care of 2-mode interactions and the associated transitions. While 2-mode resonances are unavoidable under changes of a single parameter of the star (say, the evolutionary time), we should keep in mind that likewise 3-, 4-, ... mode resonances are equally unavoidable. There always exist discrete epochs τ and (positive or negative) integers n, m, k such that

$$n[\omega_a + \tau A] + m[\omega_b + \tau B] + k[\omega_c + \tau C] = 0, \qquad (24)$$

can be satisfied to any precision we wish ($\omega_a + \tau A$ represents the local evolution of mode a, etc., around the evolutionary epoch $\tau = 0$; it is assumed that $A, B, C, \neq 0$). The order of a resonance, however, typically increases as the number of resonant modes increases. Experience shows that the lowest order resonances influence the dynamics most significantly (rigorous results are known in this respect for Hamiltonian systems, cf. Arnold[12]). Therefore, one may hope that only lowest dimensional maps will be needed to analyze the relevant resonances. Whether the dimension 2, as considered here, exhausts the qualitative description of resonance-induced transitions in real stars remains doubtful, however. There are observational indications that 3-mode resonances of lowest order do occur in some of the classic variables (Beat Cepheids,[13] Dwarf Cepheids[14]). An extension of the formalism presented here to three-dimensional maps is needed to take account of the latter.

NUMERICAL EXPERIMENTS

The numerical experiments described in Paper I were concerned with the behavior of the map in sections of the originally five-dimensional parameter space (subspaces $b_I = 0, c_I = 0$). The latter remain subspaces $a_I = 0, b_I = 0, c_I = 0$ of the higher dimensional new parameter space $\triangle^6$ of the family **(23)**. An arbitrary point X in the control space $\mathbb{C}^8 = \mathbb{D}^2 \times \triangle^6$, of coordinates $(z_1, z_2, a_R, a_I, b_R, b_I, c_R, c_I)$, uniquely defines an iterative sequence

$$s = \{q(0), q(1), q(2), \ldots\}. \tag{25}$$

The carrier of the finite sequences s, denoted by $\mathbb{B}(q;\mathbf{p}) \subset \mathbb{C}^8$, is the *bounding region*. Since only the latter is physically meaningful, we identify the control space with the bounding region

$$\mathbb{B}(q;\mathbf{p}) \equiv \mathbb{C}^8 \subset \mathbb{E}^8. \tag{26}$$

The bounding region is partitioned into basins of attraction, $\mathbb{C}^8(T)$, carrying the different types T of oscillations [Eq. **(12)**]. In Paper I we have considered a coarser partition, in which the carrier of all P-K oscillations of given K ($=1, 2, \ldots$), denoted $\mathbb{B}^{(K)}(q;\mathbf{p})$, and the carrier of all chaotic oscillations, $\mathbb{B}^{(c)}(q;\mathbf{p})$ are grouped together. We thus have

$$\mathbb{B}^{(K)}(q;\mathbf{p}) = \cup_{T \in K} \mathbb{C}^8(T), \tag{27a}$$

$$\mathbb{B}^{(c)}(q;\mathbf{p}) = \cup_{T \in c} \mathbb{C}^8(T), \tag{27b}$$

where the notation $T \in K$ and $T \in c$ means that the types T range over all period-K oscillations, or over all chaotic oscillations, respectively. The intersection of the bounding region by a fixed-parameter slice, $\mathbf{p}$ constant (*finite-amplitude basin*), was denoted by $\mathbb{F}_\mathbf{p}(q)$. With our identification **(26)** the latter becomes just the behavior space $\mathbb{D} \subset \mathbb{E}^2$ of an individual map of the family, namely the map specified by a set of fixed values of the parameters $\mathbf{p}$.

$$\mathbb{D}^2 \equiv \mathbb{F}_\mathbf{p}(q) = \mathbb{B}(q;\mathbf{p}) \cap \{\mathbf{p} = \text{constant}\}. \tag{28}$$

The coarse partition of the behavior space of a map of fixed-parameter values $\mathbf{p}$ into P-K basins, $F_\mathbf{p}^{(K)}(q)$, $K = 1, 2, \ldots$, and C basins, $F_\mathbf{p}^{(c)}(q)$, carriers of the period-K oscillations, $K = 1, 2, \ldots$, and the chaotic solutions, respectively, is obtained by setting $\mathbf{p}$ constant in the coarse partition **(27)** of the control space. Finally, the *Mandelbrot* set $\mathbb{M}_q(\mathbf{p})$, defined as the intersection of the bounding region with a fixed set of initial conditions, is now identified with the allowed parameter space $\triangle^6$ (for fixed initial conditions), since outside this set the map ceases to be physically meaningful. In Paper I we have multiplied the defining functions $\mathbf{F}[\mathbf{y}]$ of our family of maps by the converging factor $1/(1 + \epsilon|\mathbf{y}|^{\alpha+2})$ (with the numerical values $\epsilon = 0.05$ and $\alpha = 2$), the motivation being that there always exists some physical mechanism preventing the amplitudes of the oscillations from growing indefinitely. Such a mechanism is indeed roughly simulated by the preceding factor. In the following numerical experiments we have included the same device. Notice that with this modification we have $\mathbb{D}^2 = \mathbb{F}_\mathbf{p}(q) \equiv \mathbb{E}^2$, $\triangle^6 = \mathbb{M}_q(\mathbf{p}) \equiv \mathbb{E}^6$, and $\mathbb{C}^8 = \mathbb{B}(q;\mathbf{p}) \equiv \mathbb{E}^8$.

A period-K oscillation is identified by the following algorithm: Given an initial condition $q(0)$, and a fixed set of basic parameters $\mathbf{p}$, we compute the sequence **(25)** up to time $T = 114$, keeping the iterates $\{q(\theta), q(\theta + 1), q(\theta + 2), \ldots, q(T - 1), q(T)\}$, $\theta = 100$. For a fixed precision $\eta = 10^{-4}$, if we can find an integer $K < (T - \theta)$

such that

$$|q(\theta) - q(\theta + \nu K)|^2 < \eta, \qquad (29)$$

for any integer ν again producing an iterate of the computed collection, then we say that the iterative sequence is a P-K oscillation. If for no integer K we satisfy inequality (29), then we term the sequence a C or *chaotic-like* oscillation. We stress that this criterion implies that what we are willing to call period-K or chaotic-like ultimately depends on the precise choice of the triplet of parameters (T, θ, η). For *any* given T and θ, we can always select an η that is small enough, so that almost no sequence of iterates will obey the precision test (29) (except those of initial conditions on a K-attractor). Alternatively, by choosing the factor η large enough, any sequence could be regarded as a P-1 oscillation. This remark has an observational counterpart. If we regard the components of $q(t)$ as a discrete time series of observed stellar parameters, the precision factor η plays the part of a measure of the observational uncertainty or noise. If this noise factor is large enough, then we will probably always be led to interpret any such signal as regular. The series will indeed be consistent with a P-K oscillation, with K some low integer, as follows from a formal application of the operational test (29). The recent controversy between Voges *et al.*[15] on the one hand and Norris and Matilsky[16] on the other regarding the question of whether the observed X-ray time series of Hercules X-1 is chaotic or not essentially reflects this undecidable situation.

Our choice of the parameter values (T, θ, η) is strongly subject to computational constraints. In order to carry out a reasonably complete survey of the coarse partition of the control space $\mathbb{C}^8$, we should take at least 100 values in the coordinates of the initial conditions, and in each of the individual six parameters, since any lower resolution gives an inadequate picture of the complicated structure of the $\mathbb{B}^{(K)}(\mathbf{q};p)$ sets. But such a resolution requires some 10^{16} iterative sequences. Even if an individual iteration takes only a fraction of a second, the time needed for a full survey of the control space would be a fraction of 10^8 years. It is quite obvious that under such circumstances one cannot afford calculating very high iterates (say, $T \geq 10^4$) in a systematic survey. For these reasons we have continued the systematic exploration of the subspace started in our previous work ($a_l = 0$, $b_l = 0$, $c_l = 0$, for the fixed initial conditions $z_1 = z_2 = 0.1$). The increase from $\theta = 50$ (Paper I) to 100, although leading to some minor changes in the details of the boundaries of the basins $\mathbb{B}^{(K)}$ in this subspace, does not alter the general fractal-like aspect of the boundaries. Test experiments in which θ was increased to 200 and 400 showed virtually no differences with the $\theta = 100$ runs. The higher precision experiments confirm the conjecture of Paper I that the boundaries $\partial\mathbb{B}^{(K)}(\mathbf{q};p)$ are fractal sets. We have also carried out a few exploratory experiments on the structure of the $\mathbb{B}^{(K)}$ basins for a complex. Again test (29) leads to fractal-like boundaries in the higher dimensional subspace ($b_l = 0$, $c_l = 0, z_1 = z_2 = 0.1$).

In the following series of numerical experiments we have analyzed the fine structure of the boundaries of the basins $\mathbb{B}^{(K)}(\mathbf{q};p)$. To this end we construct the (a,b) sections of these basins at constant c and constant initial conditions; all parameters are real, as in Paper I. In the color photographs (FIGS. 1 and 2) showing the partition of the (a,b)-plane into the various basins of different time behavior the following color code has been chosen: white areas carry P-1 (or P-9) oscillations; red areas

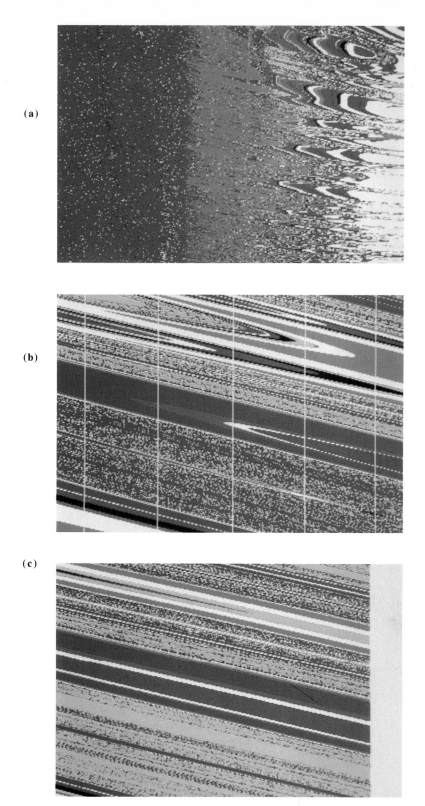

FIGURE 1. The (a,b)-plane for $c = -1.5$. (a) $-0.2 \leq a \leq 1.8$; $-1.5 \leq b \leq 1.5$. (b) $-0.03 \leq a \leq 0.01$; $0.1 \leq b \leq 0.16$. (c) $-0.015 \leq a \leq 0.011$; $-0.105 \leq b \leq 0.111$.

(a)

(b)

(c)

FIGURE 2. The (a,b)-plane for $c = -0.5$. **(a)** $-0.03 \leq a \leq 0.07$; $-1.48 \leq b \leq -1.33$. **(b)** $0.042 \leq a \leq 0.043$; $-1.362 \leq b \leq -1.3605$. **(c)** $0.0425 \leq a \leq 0.0426$; $-1.361 \leq b \leq -1.36085$.

correspond to P-2 (or P-10) solutions; blue: P-3 (or chaos-like); cyan: P-4 (or P-11); black: P-5 (or P-12); green: P-6 (or P-13); magenta: P-7; yellow: P-8. The period-1 basin has been subdivided into a basin of "stable equilibria" (painted magenta) corresponding to iterations $q(t) \to 0$ as $t \to \infty$. Physically, these solutions represent a stable nonoscillating reference state of the star; the remaining part of the period-1 basin, supporting in principle either genuine oscillations, or possibly also new equilibrium states, are all shown in white. The ambiguity in the color scheme is in part resolved by the observation that usually for low values of the stability parameter a we encounter period-K oscillations with lowest K.

FIGURE 1 displays a typical sequence of successively magnified zones of the (a,b) parameter plane ($c = -1.5$; $z_1 = z_2 = 0.1$). Frame (a) shows the partition over a broad range of the parameters ($-0.2 \le a \le 1.8$; $-1.5 \le b \le 1.5$). On a rough scale we notice a white central rectangle with a triangular roof (P-1); the white region is surrounded by a red area (P-2), which in turn is surrounded by a cyan area (P-4), and then by a finer yellow stripe (P-8); the whole picture is immersed in a blue field (C, lying outside the frame). As we increase a at fixed b (take, for instance, $b \approx 0.5$), we successively pass through Feigenbaum-like period bifurcations. Since the parameter a is a measure of the linear growth-rate, the star follows the Feigenbaum scenario as it becomes more and more vibrationally unstable. Recall that this sequence of events is in qualitative agreement with the hydrocode calculations by Kovacs and Buchler[2] for their population II stars of intermediate luminosity to mass ratios, increasing a values corresponding to decreasing effective temperatures. At low resolution the boundaries between the regions of P-2^k behavior, $k = 0, 1, 2, \ldots$, have a rather smooth appearance. The 380×250 pixel resolution of frame (a) demonstrates already a definite small-scale structure, and, besides the conspicuous P-2^k basins, it discloses the existence of basins of other types of behavior. Frame (b) is a 50-fold enlargement of the lower left corner of the white rectangle ($-0.03 \le a \le 0.01$; $0.1 \le b \le 0.16$). The P-1 and P-2 basins are now seen to interpenetrate to form a characteristic repetitive pattern of alternatively white and red bands separated by parabolic boundaries squeezed between a system of roughly parallel straight lines; the latter, at an angle of about 8° with the b-axis, reappear toward the left as cyan lines (P-4 behavior). As we continuously increase the value of a at fixed b, we alternatively pass through P-1 and P-2 stripes. Frame (b) also indicates that the chaotic basin penetrates into the P-1 and P-2 basins, forming parabolic patterns similar to the patterns drawn out by the P-2 basin. Frame (c) is finally a tenfold magnification of frame (b) ($-0.015 \le a \le 0.011$; $0.105 \le b \le 0.111$). With the exception of the neighborhood of the vertex of a parabola (cf. middle of the frame), the regions of different behavior now show up as parallel stripes whose widths seemingly tend to zero, while more and more stripes appear as we approach the straight lines of frame (b). These experiments provide evidence that the boundary between the red and white, the red and blue, and the white and blue zones in the (a,b) parameter plane has a length that increases and diverges as it is estimated at increasingly higher resolution [cf. frames (b) and (c)].

The boundary structures in frames (b) and (c) are not exceptional. Our experiments indicate that an interpenetration of just two or three different basins is only one simple alternative among much more complicated occurrences. In many parts of the (a,b) parameter plane large numbers of different basins are found to coexist. An illustration is provided in FIGURE 2 ($c = -0.5; z_1 = z_2 = 0.1$). Frame (a) now exhibits the partition of the (a,b)-plane over an already small parameter range ($-0.03 \leq a \leq 0.07; -1.48 \leq b \leq -1.33$). As we move from right to left in this frame at constant a, we first have an area supporting predominantly P-1 oscillations. Then we see an extremely irregular area in which all classes of oscillations coexist; notice that besides the areas supporting the familiar P-2^k oscillations, we have relatively extended magenta (P-7), black (P-5), and green (P-6) areas, as well as blue areas (here probably P-3). In the middle of the frame we observe predominantly P-4 oscillations. Finally, toward the left, two-fifths of the frame is essentially filled with C oscillations. Frame (b) is a hundredfold magnification of the range ($0.042 \leq a \leq 0.043; -1.362 \leq b \leq -1.3605$) in the irregular area of frame (a). We observe again a pattern of a structure similar to frame (c) of FIGURE 1, made up of areas supporting different types of time behavior that are separated by parabolas. Instead of the three different basins of FIGURE 1(c) we here notice the coexistence of a large number of basins. Frame (c) is a further tenfold magnification of the range ($0.0425 \leq a \leq 0.0426; -1.361 \leq b \leq -1.36085$). This frame displays parallel stripes of apparently infinitesimal widths, alternating with broader stripes of all colors.

To study the geometry of the boundary set we first compare frames (b) and (c) of FIGURE 2. The prominent stripes in frame (b), say, near the second white vertical line from the right ($b \approx -1.361$), show the following sequence of colors (Σ) as we go from higher to lower values of a: white, cyan, white, yellow, white, cyan, white, which is exactly the same sequence seen on the magnified frame (c). At lower a values we observe another conspicuous color sequence (Σ'): magenta, blue, green, white, red, white, green, blue, magenta, again both on frame (b) and on frame (c). Accordingly, at the level of resolution of frames (b) and (c), and therefore, presumably also at higher resolutions, the partition of the (a,b)-plane into regions of different time behavior possesses a *self-similar structure*. Consider next any segment transverse to the system of stripes, for instance, again the vertical line near $b \approx -0.1361$. The intersection with the stripes partitions this segment into subsegments of different colors whose hierarchical organization can be mimicked by the following construction. Start with a line segment of length L; divide this segment into $Q = 5$ (or perhaps 6) subsegments of same length, and paint the second according to the color scheme (Σ), and the fourth according to the scheme (Σ'). In the second step, repeat the same operations with each of the $P = 3$ (or 4) remaining neutral subsets of step one, etc. In the limit of an infinite number of repetitions of the same operations the remaining neutral segments shrink to the boundary between the blocks of the two color sequences, Σ and Σ'. The construction implies that this boundary is a *Cantor set* $\mathbb{K}$ of fractal dimension

$$F = \frac{\ln P}{\ln Q}. \tag{30}$$

Our specific model construction then gives $F = 0.68$ (0.77). Under the assumption that the (a,b) sections of the partition of the basins $\mathbb{B}^{(K)}$ in the control space $\mathbb{C}^8$ are generic, the boundary $\partial\mathbb{B}^{(K)}$ is a direct product of a Cantor set and a seven-dimensional smooth manifold, so that we may tentatively conclude that the fractal dimension of the boundary $\partial\mathbb{B}^{(K)}$ is equal to $7 + F \approx 7.7$ in the framework of the preceding model. The hypothesis of genericity has been confirmed by a few experiments in the (a,c)-plane and in randomly oriented planes of the subspace of real parameters. A few exploratory runs have shown that the Cantor set structure in the (a_R,b)-plane survives when $a_I \neq 0$.

CONCLUSION

The high precision survey of the parameter space carried out in this paper provides strong evidence for the fractal nature of the boundaries between basins of different time behavior. The fractal geometry turns out to be simpler than suggested by the lower resolution plots showing broad parameter ranges, such as frame (a) of FIGURE 1 (cf. also Paper I). Locally, in the neighborhood of a generic point λ of a generic 2-parameter section, these boundaries are given by a Cantor set $\mathbb{K}$ (the "neutral segments" separating the "colored segments" of our geometric construction) times a line segment $\mathbb{L}$. But the H–R diagram is a two-dimensional section of the parameter space of realistic stars, and therefore also of the parameter space $\triangle$ of our algebraic scheme. Our model then implies that parts of the H–R diagram are locally covered by a network of parallel lines that are the boundaries or regions of different oscillatory behavior. Given a star of effective temperature T and luminosity L in such a part, we have basically two distinct alternatives.

(1) Position (T,L) lies in a band passing through a "colored segment" in an early stage of the Cantor set construction. In the course of its evolution the star will then remain in a well-defined state of oscillation for some finite evolutionary period. Direct transitions between *any* two types of oscillations are possible (any alternation of colors being found in our experiments).

(2) Position (T,L) lies in a band passing through a "neutral segment" of some later stage of the Cantor set construction. The evolution then constrains the star to switch a large number of times from one oscillatory state to another, thereby remaining in a transitional state. The time behavior of such a star will appear as erratic or "chaotic," even though the sequence of oscillatory states the star traverses may not contain a genuine chaotic state. This type of chaos, conveniently referred to as a *Cantor transition,* is not captured by the hydrocode experiments, since the latter are based on equilibrium models.

At the moment it is not possible to make a reliable estimate of the probability for a variable star to get caught in a Cantor transition. To this end, at least two successive stages of the Cantor set construction need to be isolated in hydrocode experiments, so that the relation between the physical parameters of realistic star models (effective temperature, luminosity, etc.) and the formal parameters a, b, c can set up. The available partial survey of the H–R diagram[2] is too coarse to allow a construction of the latter relationship. The width of the effective temperature range over which one observes regular oscillations is in order of magnitude 10^3 K (cf. reference 2, FIG. 3), while in our iterative maps the width of the a-interval over which we have regular oscillations is of the order of 1 (cf. FIG. 1). The Cantor set structure becomes manifest over a-ranges of 10^{-3} to 10^{-4}; it therefore requires steps in the effective temperature $< 10^{-1}$ K to become detectable. The fraction of variables undergoing a

Cantor transition may be estimated as $< 10^{-4}$. Finally, the two-dimensional map shows that P-3, P-5, P-6, and P-7 oscillations have nonzero measures in the parameter space, and therefore finite probabilities of occurrence in real stars. There is now evidence that a P-3 oscillation has been observed in the ZZ Ceti variable GD66.[17,18]

ACKNOWLEDGMENTS

The author wishes to thank an unknown referee for his careful reading of an earlier version of the manuscript. He also thanks the members of the Institute of Astronomy, Cambridge, for their hospitality.

REFERENCES

1. BUCHLER, J. R. & G. KOVACS. 1987. Astrophy. J., Lett. **320:** 157.
2. KOVACS, G. & J. R. BUCHLER. 1988. Astrophys. J. **334:** 971.
3. BUCHLER, J. R. 1989. *In* NATO ARW on Numerical Modelling in Astrophysical Fluid Dynamics: 1. Les Arcs, March 20–24, 1989.
4. PERDANG, J. 1989. *In* NATO ARW on Numerical Modelling in Astrophysical Fluid Dynamics: 333. Les Arcs, March 20–24, 1989.
5. ANDRONOV, A. A., A. A. VITT & S. E. KHAIKIN. 1966. Theory of Oscillators. Pergamon. Oxford.
6. THOM, R. 1972. Stabilité Structurelle et Morphogénèse. Benjamin. Reading, Mass.
7. POSTON, T. & I. STEWART. 1978. Catastrophe Theory and Its Applications. Pitman. London.
8. PERDANG, J. 1988. Fractals and Chaos. Lecture Notes, Maîtrise en Astrophysique et Géophysique. Université de Liège, Liège, Belgium.
9. LAUWERIER, H. A. 1986. *In* Chaos, A. V. Holden, Ed.: 39, 58. Manchester University Press. Manchester, England.
10. OLVER, P. J. 1986. Application of Lie Groups to Differential Equations. Springer-Verlag. New York/Berlin.
11. ARNOLD, V. I. 1983. Geometrical Methods in the Theory of Ordinary Differential Equations. Springer-Verlag. New York/Berlin.
12. ———. 1976. Méthodes mathématiques de la mécanique classique. Mir. Moscow.
13. SIMON, N. R. 1979. Astron. Astrophy. **74:** 30.
14. ———. 1979. Astron. Astrophy. **75:** 140.
15. VOGES, W., H. ATMANSPACHER & H. SCHEINGRABER. 1987. Astrophys. J. **320:** 794.
16. NORRIS, J. P. & A. MATILSKY. 1989. Astrophys. J. **346:** 912.
17. AUVERGNE, M., A. BAGLIN & G. VAUCLAIR. 1988. *In* Advances in Helio- and Astroseismology. IAU Symposium, Vol. 123, p 339.
18. VAUCLAIR, G., M. J. GOUPIL, M. AUVERGNE & M. CHEVRETON. 1989. Astron. Astrophys. **215:** L17.

The Extended Atmospheres of Pulsating Stars[a]

G. H. BOWEN

Astronomy Program
Physics Department
Iowa State University
Ames, Iowa 50011

INTRODUCTION

There are numerous types of oscillating stars, and these have been much studied. They are interesting not merely because they present challenging puzzles to be solved, but because their oscillatory behavior gives valuable clues to the internal structure and processes of stars, and because they often play roles of great importance in stellar evolution.

Theoretical work on a given class of oscillating stars typically begins with the use of linearized equations to study the behavior of the stellar interior for oscillations of very small amplitude. In this approximation the stellar atmosphere can be regarded as an essentially passive element that rises and falls on the pulsating interior with negligible change in itself or effect on the rest of the system (see Cox,[1] p. 79). Large-amplitude pulsation is common in real stars, however, and modeling the complex nonlinear behavior that results is more difficult. Among other things, the atmosphere is then no longer a passive, unchanging part of the system. The structure of the dynamic atmosphere is likely to be very different from that of a static atmosphere, and it can play a vital part in various phenomena. In particular, stellar winds developing in the atmospheres of pulsating stars can cause mass loss with major consequences for stellar evolution.

To illustrate these effects, this paper will present some of my own work on dynamical modeling of the atmospheres of Mira variables, a very numerous class of long-period variable stars (LPVs). These are enormous, cool stars on the asymptotic giant branch (AGB), which derive their energy from hydrogen and helium burning shells around a tiny degenerate core of carbon and oxygen. Typical Mira parameters are believed to be: a mass of 1–$2M_\odot$, a radius of 200–$300R_\odot$, an effective temperature of 2800–3000 K, and a luminosity of perhaps 2500–$5000L_\odot$.[2] They undergo large-amplitude radial pulsation, generally in the fundamental mode, with periods of 200–500 days. They have relatively slow, cool outflowing winds, with typical mass loss rates of perhaps 10^{-7} to a few times 10^{-6} $M_\odot$ per year, and there is considerable circumstellar dust.

[a]This work was supported by the following NASA grants: Astrophysics Theory Program Grant NAGW-1364, IUE Grant NAG5-707, and ADP Grant NAG5-1181. The extensive numerical calculations were made possible by a generous grant from the Iowa State University Computation Center.

The Mira variables represent a normal stage in the evolution of stars of low to intermediate mass. Virtually all of the stellar envelope is removed by the wind in a time of the order of 10^6 years, and the core becomes a white dwarf with a mass of around $0.6M_\odot$. Similar mass loss processes apparently enable stars with initial masses up to at least $5-6M_\odot$, and probably up to about $8M_\odot$, to end their lives peacefully as low-mass white dwarfs. If it were not for this remarkable loss of mass, many more supernovas would occur than are actually observed. The consequences for the stellar population, for the chemical composition of the interstellar medium, and indeed for the structure of the Galaxy are profound.

Dynamical modeling calculations for Mira-like stars have led to the following picture. Interior modeling shows that the pulsation is driven by opacity changes in the hydrogen ionization zone,[3,4] whose radius is typically a little more than 0.9 of the photosphere radius. Waves traveling outward from there in gas of decreasing density grow rapidly in amplitude, and periodic shocks form at the photosphere, or not far beyond. Outside the radius at which shocks form, the dynamic structure of the atmosphere is very different from that of static models.[5-8] It becomes enormously extended, so that there is significant density even at distances of many stellar radii. Most of the energy transported by the shocks is dissipated in postshock regions in the relatively dense inner atmosphere, and the shocks weaken as they move further outward. Dust grains form in the cool outer regions of the atmosphere, are accelerated outward by radiation pressure, and transfer momentum to the gas by means of collisions. The gas thus accelerates to a speed greater than the local escape velocity, and flows outward as a fairly steady, slow, cool wind. The resulting mass loss rates for models are comparable to those observed for real stars. Let us now examine some of these results in more detail.

MODELING METHOD

The method used has been described in some detail in Bowen.[5] Shorter summaries, including some later changes and improvements, are given in Bowen.[6,8] In essence the method is quite straightforward: hydrodynamic equations are written for concentric spherical Lagrangian shells in the model, whose inner radius is placed inside the photosphere, but just outside the driving zone of the star, and whose outer radius is large enough to include all effects of importance in the atmosphere (usually at least 20 stellar radii); the inner boundary is constrained to oscillate periodically in the radial direction, thus simulating the interior pulsation of the star; and the response of the model atmosphere to this driving is found by explicit numerical integration of the equations, using standard finite-difference methods. It is then possible to study how the model's structure and behavior depend on the stellar parameters used (e.g., mass, radius, period), on the physical properties assumed (e.g., opacity, grain condensation temperature), and on the physical processes represented in the modeling equations (e.g., time-dependent relaxation phenomena, heating of the gas by collisions with grains). This has made it possible to gain much insight into the behavior of these complex stellar systems.

No constraints are placed on the form of the solution, except for those implied by the assumptions that rotation and magnetic effects can be neglected and that the pulsation is strictly radial. A number of approximations are used to bridge our ignorance or to make the calculations feasible, but there is continuing effort to improve these by introducing more basic physics and better computational methods—for example, in time-dependent chemistry, in radiative transfer, and in grain formation and dynamics—and of course to concentrate these efforts in the areas that earlier exploratory work has indicated are most critically important.

It seems appropriate to point out here that (1) neither the basic equations nor the methods of solution have been "linearized" in any way, so there has been no approximation in this sense; (2) time-dependent processes abound in these systems (e.g., radiative heating and cooling, molecule formation and dissociation, ionization and recombination, grain formation and growth), and for much of the time, in much of the atmosphere, most of these are *not* in equilibrium; and (3) interactions between different processes are very important and lead to a rich complexity in the resulting behavior. Results in the complete system are *not* simply the sum of the separate effects that might be calculated for independent processes. Nevertheless, the phenomena can be analyzed; they *can* be understood.

SOME MODELING RESULTS

FIGURES 1–4 present graphically the results of calculations for a representative model: the atmosphere of a $1.0 M_\odot$, $240 R_\odot$, 3000-K star, which has a fundamental mode period of 320 days, according to the period–mass–radius relationship of Ostlie and Cox.[9] The model was driven at its inner boundary (often called the *piston*) at the fundamental period of 320 days, with a velocity amplitude of 3.0 km s^{-1}. The amount of dust present was adjusted so that in regions cool enough for maximum dust formation the outward acceleration that would result from radiation pressure on the grains was 0.95 g.

FIGURE 1 shows the radial velocity as a function of radius at four phases. A strong shock is formed at about the photosphere; it can be seen propagating outward through the extended atmosphere and becoming much weaker. In the region where dust begins to form there is rapid acceleration; beyond about $15 R_{star}$ the speed exceeds the escape velocity, and a nearly constant wind velocity of 11–12 km^{-1} is reached. The wind has been formed.

FIGURE 2 shows the gas kinetic temperature as a function of radius at four phases. In the relatively dense inner atmosphere collisional excitation, and hence also radiative cooling are rapid, so that the postshock region of elevated temperature is a very narrow spike in the plot. At larger radius the density becomes low, and postshock cooling is then much slower. In fact, the expansion between shocks becomes effectively adiabatic, and the temperature in regions between shocks can drop below the radiative equilibrium temperature (TEQ) there. At still larger radii, where the outflow is almost steady and grains are fully formed, the temperature varies little with time, but remains slightly above TEQ because of heating by collisions with grains.

FIGURE 3 shows the density as a function of radius. Note that the density inside the shock formation region is almost the same as that in a static atmosphere, decreasing exponentially with a scale height that is about 0.025 times the photo-

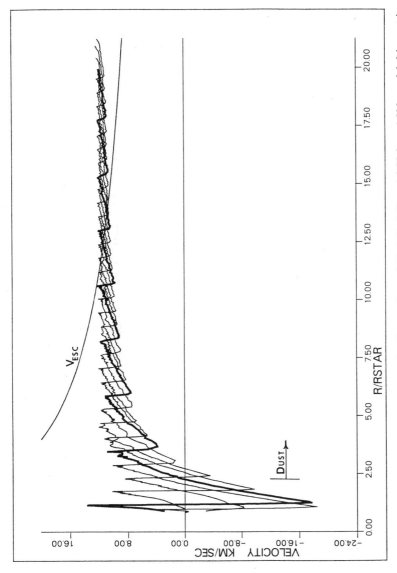

FIGURE 1. Radial velocity as a function of radius at phases 0.00, 0.25, 0.50, and 0.75 for a $1.0M_\odot$ model driven at its fundamental mode period of 320 days with a piston velocity amplitude of 3.0 km s^{-1}. (Phase 0.0 is shown *bold.*) Dust is present.

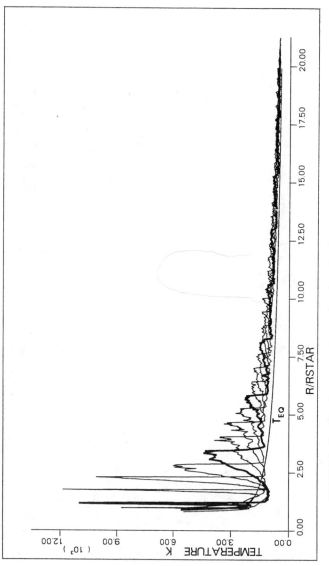

FIGURE 2. Gas kinetic temperature as a function of radius for the same model as for FIGURE 1.

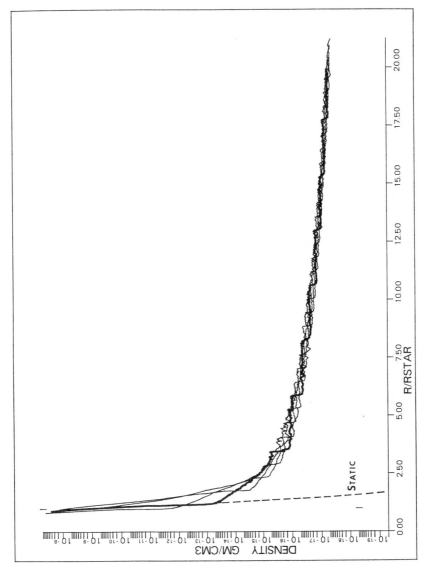

FIGURE 3. Density as a function of radius for the same model as for FIGURE 1.

sphere radius. At larger radii, however, the dynamic atmosphere is strikingly different. The density there is low, but remains significant out to very large distances. This is the circumstellar wind region, of course, where the radial velocity is approximately constant; the density there is proportional to r^{-2}, as it must be for constant mass flow at constant velocity.

FIGURE 4 shows the radius of selected Lagrangian shells (by no means all) as a function of time through two complete pulsation cycles. This shows only the inner, relatively dense region, where the average outward velocity is small; at larger radii the formation of an almost steady outflowing wind becomes obvious, but a graph scaled to show that obscures many interesting features of the inner region. Note, by the way, that the total excursion of a given parcel of material here can be as much as 60 percent of the stellar radius. This is *not* a small amplitude!

In addition to the shock formed near the photosphere at about model phase zero, when the piston is moving outward, there is a second shock formed at about phase

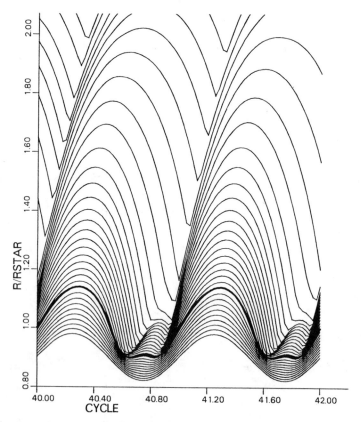

FIGURE 4. Radius of selected shells as a function of phase for the same model as for FIGURES 1–3. The *bold line* is the photosphere; the *line at smallest radius* is the piston. Piston velocity amplitude was 3.0 km s^{-1}.

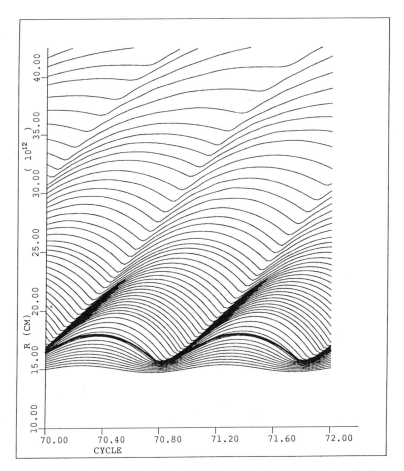

FIGURE 5. Radius of selected shells as a function of phase for the same model as for FIGURE 4. Driving at the overtone period of 152 days with a piston velocity amplitude of 1.5 km s^{-1}.

0.6. This "intraperiod" shock looks rather insignificant, but it occurs in a region of relatively high density and for a short time actually dissipates considerably greater power than the "main" shock.

This intraperiod shock is not an artifact of the modeling procedure. Its cause has been traced to the existence of an acoustic cutoff period for wave propagation.[8] It has long been known (Lamb,[10] sec. 309) that sinusoidal waves cannot propagate radially outward in the gravitationally established density gradient of an atmosphere if their period is greater than $(4\pi H/c_s)$, where H is the density scale height and c_s is the sound speed. For the model of FIGURES 1–4 the fundamental period is 320 days, and the first overtone period is 152 days. The acoustic cut-off period is a function of radius, going through a rather broad minimum at a value of about 135 days just inside the photosphere.

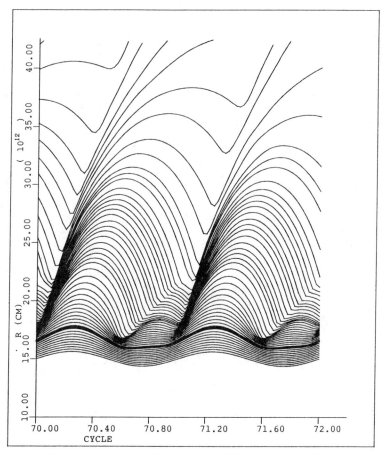

FIGURE 6. Same as FIGURE 5, but driven at the fundamental period of 320 days with a velocity amplitude of 1.5 km s⁻¹.

FIGURES 5–7 show the strikingly different results obtained with driving periods of 152, 320, and 410 days. We can understand this as follows. When the model is driven at the fundamental period, as in FIGURE 6, most of the acoustic power is reflected back into the star, and an almost pure standing wave is formed inside the photosphere. Because the waves traveling outward are not exactly sinusoidal, some power is present in components of shorter periods that do pass through the reflecting region and grow in amplitude. Material lofted by these waves does not remain "levitated" for the full 320-day period, however; it falls inward, producing an intraperiod shock before the next main shock develops. The brief burst of strong energy dissipation in the intraperiod shock may be responsible for the "bump" often seen in LPV light curves.[11]

When the model is driven at the first overtone period, there is little reflection. The waves inside the photosphere are almost pure traveling waves. The pattern seen in FIGURE 5 is accordingly very simple. No intraperiod shock is formed at this or any other driving amplitude. With a driving period of 410 days (about three times the cutoff value) *two* intraperiod shocks are formed, as seen in FIGURE 7. And using other models, it has been shown that the crucial driving periods for behavior like that shown in FIGURES 5–7 scale with the star's mass and radius exactly as does the predicted cutoff period.

This behavior is not just a curiosity. It has a corollary that appears to be significant. Acoustic power transmission into the atmosphere, where it is dissipated, is strongly frequency dependent, as shown in FIGURE 8. To sustain large-amplitude pulsation in the overtone mode requires a very large acoustic power input, easily reaching a substantial fraction of the entire stellar luminosity in the models. The average power required for fundamental mode pulsation of the same amplitude is much, much less. In the models the piston can deliver any power we may request, of

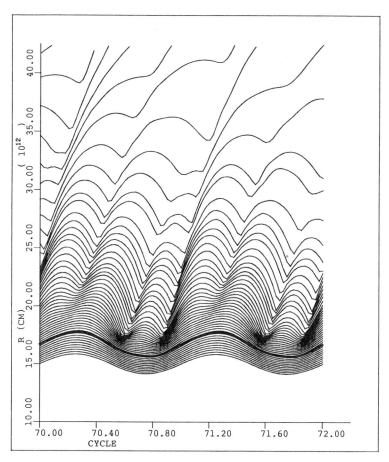

FIGURE 7. Same as FIGURES 4 and 5, but driven at a period of 410 days with a velocity amplitude of 1.5 km s^{-1}.

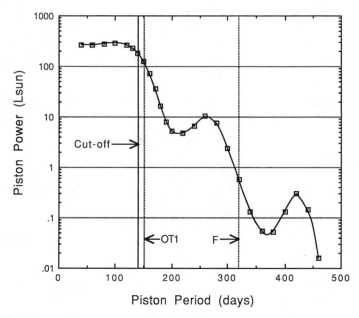

FIGURE 8. Average piston power as a function of piston period at a piston velocity amplitude of 1.5 km s^{-1}. Same model as for FIGURES 5–7.

course, but the driving zone in real stars cannot deliver unlimited power. This must limit the maximum possible amplitude for overtone mode pulsation to much smaller values than for the fundamental mode. In effect, this strongly favors the fundamental as the usual pulsation mode in such stars.

MASS LOSS AND LPV EVOLUTION

It appears that stars leave the AGB with a rather narrow distribution of masses that is strongly peaked in the vicinity of $0.6M_{\odot}$.[12] The reason for this has remained a puzzle, however. The empirical expression for mass loss rates suggested by Reimers[13] has been widely used in conjunction with some kinds of evolutionary calculations, but it does not turn on rapidly enough, nor does it reach high enough values, to be satisfactory for this purpose—and in any case, it does not elucidate the physical mechanism that must be involved. The term *superwind* has been suggested for the rapidly developing and very strong wind that clearly seems needed and there has been considerable speculation about its nature and origin.[14–16]

Modeling results for LPV atmospheres appear to offer a satisfactory answer to this problem. It became clear some time ago[5] that if one compares a series of models that all have the same mass, but that have progressively larger and larger radii, hence longer pulsation periods, there is a striking increase in the mass loss rate, $\dot{m}$; similarly, a series of models having the same period, but *decreasing* mass, will also show increasing $\dot{m}$. In both cases the essential factor is that the surface gravity is decreasing, which increases the atmosphere's scale height, which in turn causes rapid increases in the gas density and in the amount of dust present in the wind formation region (rapid because the scale height appears in an exponential function), and $\dot{m}$ increases very rapidly. In fact, for models with masses of the order of $1.0-1.2M_{\odot}$, the dust usually becomes optically thick for periods much greater than 500 days. (Models for more massive stars behave similarly, but at somewhat longer periods.) They then have properties resembling those of OH/IR sources, rather than optical Miras. This seems a likely reason why few Miras are identified that have periods much larger than about 500 days.

This rapid increase in mass loss rate with increasing stellar radius and pulsation period, or with decreasing mass, is exactly the kind of thing needed to give a final mass distribution like that previously described. This is best seen in a plot of $\log M$ vs. $\log L$, like FIGURE 9, in which a number of things have been shown. First, recall that as evolution proceeds, there is a continuous processing of hydrogen (and of helium, at times), so that the star's core grows steadily. There is a relationship between the star's luminosity, L, and the hydrogen-depleted core mass, M_c, which Paczynski[17]

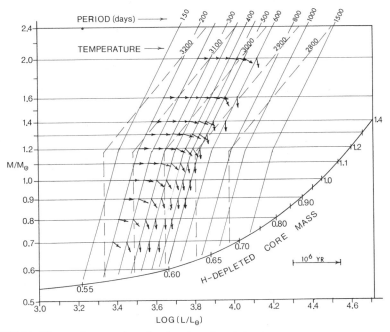

FIGURE 9. The *small arrows* show the direction of evolution in this diagram for models with the mass, luminosity, effective temperature, and fundamental mode period indicated by the various scales on the graph. All models driven at their fundamental-mode period with the amplitude that gives a peak power input from the piston equal to the star's luminosity. (The average power input is much less.) See text for further description.

gives as

$$L = 59,250(M_c - 0.522),\qquad(1)$$

where both quantities are given in solar units. FIGURE 9 includes this curve.

If there were no mass loss at all, the evolutionary track for a given star would consist of a straight horizontal line from left to right until it intersects the core mass curve and the star becomes a (rather massive) white dwarf—or the core mass exceeds $1.4M_\odot$ and a supernova occurs. If the Reimers mass loss expression holds, the track curves downward, and stars with a wider range of initial masses reach the core mass line at less than $1.4M_\odot$; the distribution of final masses is not at all like that which actually occurs, however.

Now consider the results with atmospheres and winds like the models I have been calculating. FIGURE 9 shows curves that indicate the effective temperatures calculated using the equation given by Iben:[18]

$$R = 312 \, (L/10^4)^{0.68}(M/1.175)^{-0.31S}(Z/0.001)^{0.088}(l/H)^{-0.52},\qquad(2)$$

where L, M, and the stellar radius R are in solar units; Z is the mass abundance of heavy elements; and (l/H) is the mixing-length-to-pressure-scale height ratio; $S = 0$ for $M \le 1.175$; and $S = 1$ otherwise. It is also necessary to use the definition of effective temperature, which gives, in solar units,

$$L = R^2(T_{eff}/5770)^4.\qquad(3)$$

The curves in FIGURE 9 were calculated using $Z = 0.020$ and $(l/H) = 0.90$.

Finally, the periods for fundamental-mode pulsation were calculated using the expression given by Ostlie and Cox[9]

$$\log P_0 = -1.92 - 0.73 \log M + 1.86 \log R,\qquad(4)$$

where P_0 is given in days and the other quantities in solar units, as before. These values are also shown in FIGURE 9.

Data were calculated for a grid of dynamic atmosphere models, with masses from 0.7 to $2.0M_\odot$ and periods from 200 to 600 days, with a series of increasing piston velocity amplitudes for each. To systematize the choice of a representative driving amplitude, the arbitrary but fairly plausible guess was made that the limiting pulsation amplitude for the star would correspond approximately to conditions when the peak piston power during the cycle equals the stellar luminosity. (The *average* piston power is much lower than that for fundamental mode pulsation.) The corresponding mass loss rate for each model was thus determined.

FIGURE 9 shows, by means of small arrows, the *direction* of evolution (not the rate) for each model. It is easy to show from Eq. (1), plus the efficiency of conversion of mass into energy, that in this kind of plot the point representing the star moves at a uniform rate of 0.245 units to the right per million years, anywhere in the diagram, as shown in FIGURE 9. Vertical motion is simply $d(\log M)/dt = 0.434 \, (\dot{M}/M)$, as determined from the modeling calculations.

The results are striking: evolutionary tracks plunge precipitously downward within a fairly short horizontal distance in the figure, or a correspondingly short evolutionary time. Stars with initial masses up to at least $2.5M_\odot$ must leave the AGB with final masses in about the right range, as previously discussed. How reliable is this result? There are a number of assumptions, approximations, and estimated modeling parameters that have gone into the calculations, of course, so that the resulting numbers have considerable uncertainty. Nevertheless, the qualitative behavior seems exactly what is needed, and many exploratory calculations have shown that the region of the M–L diagram where it occurs is surprisingly insensitive to most changes in parameters. Work is continuing on this, but there is reason to be optimistic that we are at least getting close to the correct answer—and an understanding of the physical mechanism for it.

CONCLUSION

It is proving to be possible to analyze successfully these very complex nonlinear systems by simply writing the fundamental equations describing the physical processes involved, and solving them by explicit numerical integration—provided one uses sufficient care! Moreover, the physical significance of the results is of very great significance for stellar evolution, for the character of the stellar population, the number of supernovas, the chemical evolution, and, in fact, the structure of galaxies.

REFERENCES

1. Cox, J. P. 1980. Theory of Stellar Pulsation. Princeton University Press. Princeton, N.J.
2. Willson, L. A. 1982. *In* Pulsations in Classical and Cataclysmic Variable Stars, J. P. Cox and C. J. Hansen, Eds.: 269–283. Joint Institute for Laboratory Astrophysics. Boulder, Colo.
3. Wood, P. R. 1974. Astrophys. J. **190:** 609–630.
4. Ostlie, D. A. 1982. Ph.D. thesis, Iowa State University, Ames.
5. Bowen, G. H. 1988. Astrophys. J. **329:** 299–317.
6. ———. 1988. *In* Pulsation and Mass Loss in Stars, R. Stalio and L. A. Willson, Eds.: 3–20. Kluwer. Dordrecht, The Netherlands.
7. ———. 1989. *In* IAU Colloquium 106: Evolution of Peculiar Red Giant Stars, H. R. Johnson and B. Zuckerman, Eds. Cambridge University Press. Cambridge, England.
8. ———. 1990. *In* The Numerical Modelling of Nonlinear Stellar Pulsations: Problems and Prospects, J. R. Buchler, Ed.: 155–171. Kluwer. Dordrecht, The Netherlands.
9. Ostlie, D. A. & A. N. Cox. 1986. Astrophys. J. **311:** 864–872.
10. Lamb, H. 1945. Hydrodynamics. Dover, New York.
11. Beach, T. E. 1990. Ph.D. thesis, Iowa State University, Ames.
12. Weidemann. 1987. *In* Late Stages of Stellar Evolution, S. Kwok and S. R. Pottasch, Eds.: 347–350. Reidel. Dordrecht, The Netherlands.
13. Reimers, D. 1975. *In* Problems in Stellar Atmospheres and Envelopes, B. Baschek, W. H. Kegel, and G. Traving, Eds.: 229. Springer-Verlag. Berlin/New York.
14. Renzini, A. 1981. *In* Physical Processes in Red Giants, I. Iben, Jr., and A. Renzini, Eds.: 165–172. Reidel. Dordrecht, The Netherlands.
15. Iben, I., Jr. & A. Renzini. 1983. Annu. Rev. Astron. Astrophys. **21:** 271–342.
16. Iben, I., Jr. 1987. *In* Late Stages of Stellar Evolution, S. Kwok and S. R. Pottasch, Eds.: 175–195. Reidel. Dordrecht, The Netherlands.
17. Paczynski, B. 1970. Acta Astron. **20:** 47.
18. Iben, I., Jr. 1984. Astrophys. J. **277:** 333–354.

Nonlinear MHD Models of Astrophysical Winds

K. TSINGANOS[a] AND E. TRUSSONI[b]

[a]*Department of Physics*
University of Crete and Research Center of Crete
Heraklion, Crete, Greece

[b]*Osservatorio Astronomico di Torino*
Pino Torinese, Italy

INTRODUCTION

The solar wind plays an important and unifying role in modern research in astrophysics and space physics. First, a detailed study of several observed and well-known solar, geophysical, and astrophysical phenomena relies on the solar wind. And second, its intrinsic physical interest as a nonlinear phenomenon provides an extra motivation for its better understanding. Parker's[1] classic theory based primarily on a spherically symmetric, hydrodynamic, and polytropic model, was further extended to include the effects of heat conduction, Alfven waves, radiation, etc., in the acceleration of the wind.[2-4] Recent observations of other wind-type astrophysical outflows from protostellar objects,[5] galactic objects like SS433,[6] and extragalactic jets,[7] however, have emphasized the need to extend the theory to nonspherically symmetric, magnetohydrodynamic, and nonpolytropic modeling. In particular, it would be very useful to have solutions of the governing MHD equations in closed analytical forms. Such analytic solutions may be used, to "zeroth order," in the testing and physical interpretation of numerical codes.[8] The hydromagnetic solutions that are presented in this paper are based on an "analytical code" that has been developed to solve the set of the nonlinear and coupled partial hydromagnetic equations.[9-11]

COLLIMATED HYDROMAGNETIC OUTFLOWS

The formal problem we address is to find solutions to the ideal MHD equations for conservation of mass, magnetic flux and momentum, and Faraday's law of induction,

$$\nabla \cdot \mathbf{B} = 0, \qquad \nabla \cdot (\rho \mathbf{V}) = 0, \tag{1a}$$

$$\nabla \times (\mathbf{V} \times \mathbf{B}) = 0, \tag{1b}$$

$$\rho(\mathbf{V} \cdot \nabla)\mathbf{V} = -\nabla P + \frac{(\nabla \times \mathbf{B}) \times \mathbf{B}}{4\pi} - \frac{\rho GM}{R^2} e_{\mathbf{R}} \tag{1c}$$

for the bulk flow speed $\mathbf{V}(R, \theta)$, magnetic field $\mathbf{B}(R, \theta)$, density $\rho(R, \theta)$, and pressure $P(R, \theta)$, in spherical coordinates R, θ, ϕ. The system of equations (1) should be

118

supplemented by an equation for the energy transport. Since the detailed energy exchange mechanisms that operate along the flow are largely unknown, however, the usual practice is to close mathematically this system via a polytropic relationship,

$$\frac{P}{P_0} = \left[\frac{\rho}{\rho_0}\right]^{\Gamma} \quad \text{or} \quad \Gamma = \frac{d \ln P}{d \ln \rho} = \text{constant}, \tag{1d}$$

with $1 \leq \Gamma < \frac{5}{3}$. Effectively, this results in specifying the distribution of the heating/cooling mechanisms along the flow in such a way as to correspond to a constant value of Γ. Then we may solve Eqs. (1) for the shape of the streamlines. Since this *a priori* specification of a constant Γ may correspond to quite an artificial heating/cooling distribution along the flow, as well as on streamline shape, we shall follow the opposite approach, that is, specify the shape of the streamlines, as it is inferred from observations of collimated outflows, for example, and then deduce from energy conservation the heating/cooling required to support such a flow pattern,

$$3(k\rho/m)(\mathbf{V} \cdot \nabla)T - 2(kT/m)(\mathbf{V} \cdot \nabla)\rho = \rho\sigma, \tag{1e}$$

$$P = \frac{2k}{m} \rho T, \tag{1f}$$

where $\rho\sigma$ is the rate of energy addition per unit volume of the fluid. The resulting value of $\Gamma(R, \theta)$ can then be compared to some characteristic values, such as $\Gamma = 1$ (isothermal atmosphere), or $\Gamma = \frac{3}{2}$ (Parker polytrope).

We search for solutions where the hydromagnetic field is helicoidal without meridional components and the dependence of the physical parameters on R and θ is separable,

$$\mathbf{V}(R, \theta) = V_R(R, \theta)e_\mathbf{R} + V_\phi(R, \theta)e_\phi, \quad V_\theta \equiv 0, \tag{2a}$$

$$\mathbf{B}(R, \theta) = B_R(R, \theta)e_\mathbf{R} + B_\phi(R, \theta)e_\phi, \quad B_\theta \equiv 0, \tag{2b}$$

$$\rho(R, \theta) = \rho(R)\rho(\theta) \tag{2c}$$

$$P(R, \theta) = P_0(R) + P_1(R) \sin^2\theta, \tag{2d}$$

$$\sigma(R, \theta) = [\sigma_0(R) + \sigma_1(R) \sin^2\theta]\sigma_2(\theta), \tag{2e}$$

The simplest θ-dependence that leads to physically interesting solutions is the following[10,11]

$$V_R(R, \theta) = V_0 Y(R) \frac{\cos \theta}{[1 + \omega \sin^2\theta]^{1/2}}, \quad B_R = \frac{B_0}{R^2} \cos \theta, \tag{3a}$$

$$V_\phi(R, \theta) = \lambda V_0 \frac{R \sin \theta}{[1 + \omega \sin^2\theta]^{1/2}} \frac{Y_* - Y(R)}{1 - M_A^2}, \quad B_\phi(R, \theta) = -\lambda B_0 \frac{\sin \theta}{R} \frac{1 - R^2/R_*^2}{1 - M_A^2}, \tag{3b}$$

$$\rho(R, \theta) = \frac{\rho_0}{Y(R)R^2}[1 + \omega \sin^2\theta], \quad \sigma_2(\theta) = \frac{\cos \theta}{[1 + \omega \sin^2\theta]^{3/2}}, \quad M_A(R) = \frac{V_R\sqrt{4\pi\rho}}{B_R}, \tag{3c}$$

where at $R = R_*$, the Alfven number $M_A(R)$ associated with the poloidal field, and flow equals unity. As will be shown in the next section, this Alfvenic point at $R = R_*$ is mathematically a sink-type of singularity.

The angular dependence of the radial flow speed, $f(\theta; \omega) = V_R(R, \theta)/[V_0 Y(R)]$, is chosen such that a single parameter ω controls the degree of collimation of the outflow,

$$f(\theta; \omega) = \frac{\cos \theta}{[1 + \omega \sin^2\theta]^{1/2}}. \tag{4}$$

For large ω the flow is strongly collimated around the polar axis, $\theta = 0$, while for smaller ω, V_R goes to zero sinusoidally with the latitude at $\theta = \pi/2$. The density increases away from the flow axis in a manner controlled similarly by the same parameter ω, to simulate the presence of a denser disk of gas at nearly static equilibrium conditions around the equator $\theta = \pi/2$. The other parameters that enter into our model are the ratio of the base radial flow speed V_0 to the base Alfven speed V_0^A, η; the ratio of the azimuthal and radial flow speeds at the base, λ; and the ratio of the escape speed and the base Alfven speed, ν,

$$V_0^A = \frac{B_0}{[4\pi\rho_0]^{1/2}}, \qquad \eta = \frac{V_0}{V_0^A}, \qquad \lambda = \frac{V_1}{V_0}, \qquad \nu = \frac{V_{esc}}{V_0^A}, \tag{5}$$

where the base Alfven speed V_0^A is associated with the poloidal magnetic field only.

With expressions (3) the R- and θ-components of the momentum balance equation (1c) yield the following ordinary differential equations for $Y(R)$, $Q_0(R) = 8\pi P_0(R)/B_0^2$, $Q_1(R) = 8\pi P_1(R)/B_0^2$,

$$Q_1 = \frac{1}{R^4} + \frac{\eta^2\lambda^2}{Y}\left[\frac{Y - Y_*}{1 - \eta^2 YR^2}\right]^2 - \frac{2\lambda^2}{R^2}\left[\frac{1 - R^2/R_*^2}{1 - \eta^2 YR^2}\right]^2, \tag{6a}$$

$$\frac{dQ_1}{dR} + \frac{\omega\nu^2}{YR^4} - \frac{2\eta^2}{R^2}\frac{dY}{dR} - \frac{2\eta^2\lambda^2}{YR}\left[\frac{Y - Y_*}{1 - \eta^2 YR^2}\right]^2 + \frac{\lambda^2}{R^2}\frac{d}{dR}\left[\frac{1 - R^2/R_*^2}{1 - \eta^2 YR^2}\right]^2 = 0, \tag{6b}$$

$$\frac{dQ_0}{dR} = \frac{\nu^2}{YR^4} + \frac{2\eta^2}{R^2}\frac{dY}{dR} = 0. \tag{6c}$$

The total angular momentum per unit mass carried by the wind, that is, the sum of the ordinary fluid angular momentum per unit mass $L = r \sin \theta V_\phi$ and the torque associated with the magnetic stresses, $L_m = r \sin \theta\, B_R B_\phi/(4\pi\rho V_R)$, is constant on each streamline θ-constant. In particular, we have chosen the following simple dependence of $L(\theta)$,

$$L(\theta) = \frac{\lambda V_0 Y_* r_*^2}{r_0} \frac{\sin^2\theta}{(1 + \omega \sin^2\theta)^{1/2}} = \Omega(\theta) r_*^2 \sin^2\theta. \tag{7}$$

Note that since near the base $r \doteq r_0$, $M_A \ll 1$ and $Y_* \gg 1$, $\Omega(\theta)$ is approximately the angular velocity of the roots of the helical field lines at the base of the wind,

$$V_\phi(R = 1, \theta) = \lambda V_0 \frac{Y_* - 1}{1 - \eta^2} \frac{\sin \theta}{\sqrt{1 + \omega \sin^2\theta}} = \Omega(\theta) r_0 \sin \theta. \tag{8}$$

Near the base then most of the angular momentum is due to the torque exerted by the magnetic field,

$$L = L_v^0 + L_m^0 \approx \Omega(\theta) r_0^2 \sin^2\theta \left[1 + R_*^2 \right], \qquad (9a)$$

with the second magnetic term dominating, while the angular velocity is

$$V_\phi(R, \theta) \approx \Omega(\theta) r \sin \theta. \qquad (9b)$$

On the other hand, further away, say at $R \gg R_*$,

$$L = L_V^\infty + L_m^\infty \approx \Omega r_*^2 \sin^2\theta \left[1 - Y_*/Y_\infty \right] \qquad (10a)$$

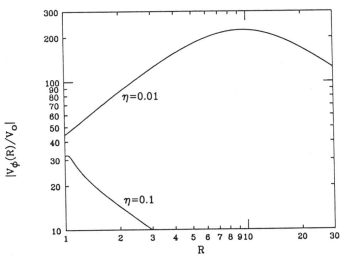

FIGURE 1. Plot of the radial dependence of the azimuthal speed V_ϕ for $\omega = 4$, $\lambda = 0.5$, $V_0 = 5$ km/s, and $\eta = 0.1$, $\eta = 0.01$. In the $\eta = 0.1$ case we have corotation up to $R_* \cong 1.24$, while in the $\eta = 0.01$ case, we have corotation up to $R_* \cong 10.55$.

and

$$V_\phi \approx r_0 \Omega(\theta) \frac{R_*^2}{R} \left[1 - \frac{V_R(R_*)}{V_R(\infty)} \right]. \qquad (10b)$$

The flow and field are approximately corotating with the angular velocity $\Omega(\theta)$ inside the Alfvenic radius R_*, $R < R_*$, while outside R_*, $R > R_*$, the angular velocity drops inversely with the distance, to conserve angular momentum. In FIGURE 1 we plot the radial dependence of the azimuthal flow speed $V_0(R, \theta = \pi/2)$ for a strongly magnetically dominated case, $\eta = 0.01$ and $R_* \approx 10.5$, and a mildly magnetically dominated case, $\eta = 0.1$ and $R_* = 1.247$.

By eliminating Q_1 between Eqs. **(6a)** and **(6b)**, we obtain a single differential equation for $Y(R)$,

$$\frac{dY}{dR} = \frac{f(R; \omega v^2; \lambda; \eta)}{g(R; \omega v^2; \lambda; \eta)}, \qquad (11a)$$

where

$$f = \frac{4}{R^5} - \frac{\omega v^2}{YR^4} - \frac{2\eta^2\lambda^2 Y}{RM^2(1-M^2)^2}\left[(2M^2-1)M^2\frac{Y_*^2}{Y^2} - (3M^2-2)\right], \tag{11b}$$

$$g = -\frac{2\eta^4 Y}{M^2} + \frac{\eta^2\lambda^2}{[1-M^2]^2}\left[\frac{Y_*^2}{Y^2}(2M^2-1) - 1\right]. \tag{11c}$$

Critical points in the differential equations that govern wind-type outflows are well known to exist; they determine the characteristic solution that meets the boundary conditions of the wind. Thus, in polytropic hydromagnetic flows,[12] there exist three critical points where the poloidal speed of the wind equals one of the characteristic speeds for the propagation of MHD waves into the medium, that is, the slow, V_s, Alfven, V_A, and fast V_f mode wave speeds. Our differential equation (11) has the Alfven critical point at R_*, where the poloidal flow speed equals to the poloidal Alfven speed, as expected. The locations where V_p equals V_s or V_f are not critical, however, because in our nonpolytropic formulation the sound speed is ill-defined, with the consequence that the fast and slow mode MHD wave speeds are not properly defined, too. The Alfven speed is independent of the equation of state used with the result that the corresponding critical point R_* appears in the final equation (11), while critical points in (11) where $V_p = V_s$, or $V_p = V_f$ do not exist. The situation is similar to the equivalent hydrodynamic solution[11] where the familiar sonic point where $V_p = c$ ($c^2 = \Gamma dP/d\rho$, $\Gamma = \frac{5}{3}$, the adiabatic sound speed) is not a critical point in the Mach number equation.[1] An analysis of the topologies of Eq. (11), however, shows that there exists a second critical point at R_x, where $f(R_x; \omega v^2; \lambda; \eta) = g(R_x; \omega v^2; \lambda; \eta) = 0$. The sole purpose of this X-type critical point is to select a single solution out of the infinite ones that pass through the Alfven critical point at R_*.

TOPOLOGIES OF THE SOLUTIONS

In FIGURE 2 we present a sample of the topology of the amplitude $Y(R)$ of the radial flow speed as η decreases from $\eta \to \infty$ (purely hydrodynamic flow), to $\eta = 1$ (equipartition between magnetic and kinetic energy), and further down to smaller values of η corresponding to increasingly magnetically dominated flows. Thus, in FIGURE 2(a) there exists a single wind-type solution with zero pressure at infinity, "breeze"-type solutions with finite asymptotic speeds (lower part of plot), and solutions that terminate at some finite distance (upper part of plot).[13] As the strength of the magnetic field increases, $\eta = 3$, a pair of critical points at R_* and R_x appears below the surface of the star at $R = 1$. The topology of the solution in the exterior space, $R > 1$, however, is essentially identical to the previous hydromagnetic case. In these hydrodynamically dominated cases, the outflow is super-Alfvenic from the base $R = 1$ to infinity. The parameters needed to completely specify the solution are five: ω, v, η, λ, and R_*.

Next consider what happens when the initial flow speed at the base is sub-Alfvenic, $\eta < 1$. In FIGURE 2(b) and (c) we show the topology of the solutions for mildly magnetically dominated cases, that is, $\eta = 0.5$ and 0.1. The pair of critical points (R_*, R_x) now appears just outside the base $R = 1$. The free parameters that

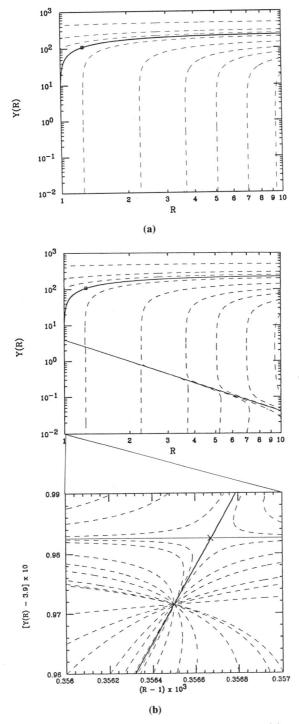

FIGURE 2. Sequence of plots of the topology of the radial dependence of the radial flow speed $V_R(R, \theta)$ for $\omega = 4$, $\lambda = 0.5$, and $V_0 = 5 \text{ km s}^{-1}$ by increasing the strength of the magnetic field at the base from a pure hydrodynamic outflow without magnetic fields in **(a)**, where $\eta = \infty$, to a strongly magnetically dominated outflow in **(d)**, where $\eta = 0.01$, and $R_* = 10.55$, $R_x = 21.86$. The intermediate cases **(b)** and **(c)** have the two critical points close to each other and to the base at $R = 1$, that is, $\eta = 0.5$, $R_* = 1.0003565$, $R_x = 1.0003568$ in **(b)**, and $\eta = 0.1$, $R_* = 1.24$, $R_x = 1.28$ in **(c)**.

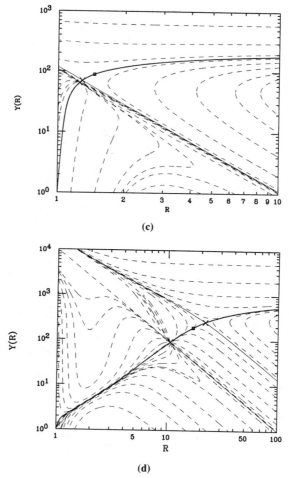

(c)

(d)

◀ *Reader: See caption on previous page.*

specify the single wind-type solution have reduced to four, because one of them, say R_*, is determined by the requirement that the solution crosses the X-type critical point and then goes to infinity with finite magnitude.

In FIGURE 2(d) the flow is magnetically dominated up to the corotation radius at $R_* = 10.5$. The radial flow speed exceeds the local Alfven speed associated with the radial component of the magnetic field at $R > R_*$. The radial flow speed exceeds the local sound speed at $R_s \cong 17$. The Alfven critical point being a singularity of the sink-type is not able to select a characteristic solution. The existence of the X-type critical point at $R_x \cong 22$ has that specific purpose: to choose out of the infinite number of curves that pass through the Alfvenic point at R_* only one, the wind solution. The other characteristic solution that crosses the X-type critical point separates "breeze"-type solutions from "accretion"-type solutions as in the classical Parker picture.[1]

REFERENCES

1. PARKER, E. N. 1958. Astrophys. J. **128:** 664.
2. HOLZER, T. H. 1977. J. Geophys. Res. **82**(A4): 23.
3. HOLLWEG, J. V. 1986. J. Geophys. Res. **91**(A4): 4111.
4. CASTOR, J. I., D. C. ABBOTT & R. I. KLEIN. Astrophys. J. **195:** 157.
5. SILVESTRO, G., A. FERRARI, R. ROSNER, E. TRUSSONI & K. TSINGANOS. 1985. Nature **325:** 246.
6. BODO, G., A. FERRARI, S. MASSAGLIA & K. TSINGANOS. 1985. Astron. Astrophys. **14:** 246.
7. FERRARI, A., E. TRUSSONI, R. ROSNER & K. TSINGANOS. 1985. Astrophys. J. **294:** 397.
8. NORMAN, M. L. 1990. Fluid dynamics of astrophysical jets. This issue.
9. TSINGANOS, K. 1982. Astrophys. J. **252:** 775.
10. HU, Y. Q. & B. C. LOW. 1989. Astrophys. J. **342:** 1049.
11. TSINGANOS, K. & E. TRUSSONI. 1991. Astron. Astrophys. In press.
12. WEBER, E. J. & L. DAVIES. 1967. Astrophys. J. **148:** 271.
13. TSINGANOS, K. & E. TRUSSONI. 1990. Astron. Astrophys. **231:** 270.

Elemental Mixing in Classical Nova Systems[a]

MARIO LIVIO[b,c] AND JAMES W. TRURAN[b]

[b]Department of Astronomy
University of Illinois
Urbana, Illinois 61801

[c]Department of Physics
Technion
Haifa 32000, Israel

INTRODUCTION

Elemental abundances play a crucial role in defining the nature of the outbursts of the classical novae.[1] Such outbursts are now known to result from thermonuclear runaways proceeding in accreted hydrogen-rich shells on the white dwarf components of close binary systems. Nuclear burning in this environment proceeds by means of the carbon–nitrogen–oxygen (CNO) -cycle hydrogen-burning reaction sequences. The energetics of the critical stages of these runaways are constrained by an important characteristic of these CNO burning sequences: the rate of nuclear energy generation at high temperatures is limited by the timescales of the slower and temperature-insensitive positron decays of ^{14}O and ^{15}O. For this reason, it has been found that the occurrence of "fast" novae—those characterized by rapid spectral development, high velocities, and super-Eddington luminosities at visual maximum[2]—demands high concentrations of carbon or oxygen nuclei in the burning shell.

Recently, elemental abundance data have become available for a number of classical nova (CN) systems (for reviews, see, e.g., Truran;[3] Truran and Livio;[4] Williams[5]). The abundance data concerning hydrogen, helium, and some heavy elements are summarized in the accompanying tables. TABLE 1 presents, specifically, the mass fractions (where known) in the form of the elements hydrogen, helium, carbon, nitrogen, oxygen, neon, sodium, magnesium, aluminum, silicon, sulphur, and iron, adapted from the indicated references. TABLE 2 provides helium-to-hydrogen ratios for a somewhat larger sample of novae. Here again, where known, the total mass fractions Z in the form of heavy elements are also tabulated. For purposes of comparison, we note that solar system matter[6,7] is characterized by a helium-to-hydrogen ratio by number He/H = 0.08 and a heavy element mass fraction $Z_{SOLAR} = 0.019$. The column labeled "enriched fraction" in TABLE 2 gives the total mass fraction of enriched matter in the form of both helium and heavy elements.

What becomes immediately apparent from these tables is the fact that *all classical novae for which relatively reliable abundance determinations exist show enrichments*

[a]This work was supported in part by National Science Foundation Grant AST 86-115000 at the University of Illinois, and in part by the Fund for the Promotion of Research at the Technion.

126

TABLE 1. Heavy-Element Abundances in Novae

Object	Year	Reference	H	He	C	N	O	Ne	Na	Mg	Al	Si	S	Fe
RR Pic	1925	34	0.53	0.43	0.0039	0.022	0.0058	0.011						
HR Del	1967	35	0.45	0.48		0.027	0.047	0.0030						
T Aur	1891	36	0.47	0.40		0.079	0.051							
PW Vul	1984	37	0.54	0.28	0.032	0.11	0.038							
V1500 Cyg	1975	27	0.49	0.21	0.070	0.075	0.13	0.023						
V1668 Cyg	1978	38	0.45	0.23	0.047	0.14	0.13	0.0068						
V693 CrA	1981	39	0.29	0.32	0.0046	0.080	0.12	0.17	0.0016	0.0076	0.0043	0.0022		
GQ Mus	1983	37	0.27	0.32	0.016	0.19	0.19	0.0034		0.0014	0.00056	0.0028	0.0016	0.00047
DQ Her	1934	40	0.34	0.095	0.045	0.23	0.29			0.0067		0.0018	0.10	0.0045
V1370 Aql	1982	41	0.053	0.088	0.035	0.14	0.051	0.52						

(relative to solar composition) in either helium or heavy elements, or both. This conclusion probably remains true even when the (large) uncertainties in the abundance determinations are taken into account. This suggests that envelope enrichment is a very general phenomenon. Truran and Livio[4] have explained why it is extremely unlikely that the source of these enrichments is either the mass transfer from the secondary star or nuclear transformations accompanying the outburst. They have concluded that the abundance patterns suggest that *some fraction of the envelope has been dredged up from the underlying white dwarf.* Given the presence of large concentrations of neon and heavier elements in the ejecta of several recent novae (e.g., V693 CrA and V1370 Aql), it is interesting to note that this conclusion has the consequence that massive ONeMg white dwarfs (as opposed to CO white dwarfs) occur quite frequently in CN systems.

The important question that immediately arises is what is the mechanism that is responsible for the mixing between the accreted envelope and the white dwarf core.

TABLE 2. Helium and Heavy-Element Abundances in Novae

Nova	Year	He/H	Z	Reference	Enriched Fraction
T Aur	1891	0.21	0.13	36	0.36
RR Pic	1925	0.20	0.039	42	0.28
DQ Her	1934	0.08	0.56	42	0.55
CP Lac	1936	0.11 ± 0.02		42	0.08
RR Tel	1946	0.19		42	0.24
DK Lac	1950	0.22 ± 0.04		42	0.30
V446 Her	1960	0.19 ± 0.03		42	0.24
V553 Her	1963	0.18 ± 0.03		42	0.23
HR Del	1967	0.23 ± 0.05	0.077	42	0.35
V1500 Cyg	1975	0.11 ± 0.01	0.30	42	0.34
V1668 Cyg	1978	0.12	0.32	38	0.38
V693 CrA	1981	0.28	0.38	39	0.61
V1370 Aql	1982	0.40	0.86	41	0.93
GQ Mus	1983	0.29	0.42	37	0.64
PW Vul	1984	0.13	0.18	37	0.27
V1819 Cyg	1986	0.19		43	0.24

It is this issue that we address in the present paper. Four possible mechanisms have been suggested: (1) diffusion-induced convection, (2) shear mixing, (3) convective-overshoot-induced flame propagation, and (4) convection-induced shear mixing. In the following sections, we discuss each of these mechanisms, pointing out its strengths and weaknesses, and we attempt to identify critical observations that will help us to determine which mixing mechanism is most likely to dominate. We survey the proposed mechanisms in the next section; our summary and a discussion of critical observations are presented in the last section.

MIXING MECHANISM

We shall now review the various mixing mechanisms that have been proposed. These can be grouped into two classes: (1) models in which most of the mixing occurs prior to the thermonuclear runaway (TNR), and (2) models in which the mixing takes place only during the TNR. It is important to note that *convection plays a crucial role in all the mechanisms.* We begin by discussing the mechanisms of the first type, for which the mixing occurs prior to the TNR.

Diffusion-Induced Convection

The mechanism of diffusion-induced convection, first discussed by Prialnik and Kovetz,[8,9] operates in the following way. During the accretion phase, small amounts of hydrogen diffuse across the boundary between the white dwarf core and the hydrogen-rich accreted envelope, where a sharp composition gradient is assumed. The hydrogen that penetrates the core (assumed to be of CO or ONeMg composition) ignites in a high Z environment. The development of convection in this region subsequently mixes CO-rich material into the envelope. A similar mechanism has been proposed by Michaud, Fontaine, and Charland,[10] for the explanation of the surface compositions of white dwarfs.

The main points demonstrated by the calculations of Kovetz and Prialnik[9,11] are the following: (1) the level of enrichment of envelope matter by white dwarf material (the enriched fraction) depends strongly on the accretion rate, and (2) the enriched fraction is relatively weakly dependent on the white dwarf's mass (or luminosity). The first point is very easy to understand, since for low accretion rates the interval between outbursts (the time required to build the critical pressure to trigger a TNR) becomes longer, thus allowing diffusion more time to operate. The level of enrichment achieved is not exactly inversely proportional to $\dot{M}$, since the partial pressure forces during the diffusion process also depend on $\dot{M}$. Nevertheless, the level of enrichment does decrease significantly with increasing $\dot{M}$ (see FIG. 1). The relatively weak dependence on white dwarf mass is a consequence of the fact that the ratio of the accreted mass Δm_{acc} to the depth (in mass) of the heavy-element-rich shell in which the TNR occurs, is found numerically to be almost independent of the white dwarf mass, M_{WD}. This point may require further study.

The main strengths of the diffusion-induced convection mechanism are the following.

1. It provides a self-consistent picture that relates the accretion phase to the TNR and that is calculable (and thus testable) even in the context of the existing one-dimensional numerical codes (apart from uncertainties in the diffusion coefficients, to be discussed later).
2. It is consistent with the fact that even novae with strong magnetic fields, such as V1500 Cyg, show a significant enrichment in heavy elements (see TABLES 1 and 2), since the diffusion process is not strongly effected by the presence of the magnetic field.
3. It is consistent with the fact that the recurrent novae U Sco and V394 CrA, for which Webbink et al.[12] suggested that the outbursts are caused by a TNR, do not show enrichments in heavy elements. For this mechanism, the absence of such enrichments would be a direct consequence of the short recurrence times of the outbursts ($\sim$ tens of years), which do not allow diffusion to take place. It should be noted, however, that these two recurrent novae show an extreme and yet unexplained overabundance of helium.

The fact that the main prediction of diffusion-induced convection is the dependence of the level of enrichment on $\dot{M}$, leads us to attempt the following test. For the CNe for which abundance determinations exist, we calculate the accretion rates obtained from magnetic braking or gravitational radiation (where appropriate), using the Mestel and Spruit[13] form of the magnetic braking law (see also, Hameury et al.[14]). According to the present version of the "cyclic evolution" ("hibernation") scenario of classical novae,[15,16] this represents the average accretion rate during the interburst interval. It is important to note that accretion rates deduced from observations of CNe[17] are considerably higher than the secular mean, and thus should not be used here (see Livio,[16] for discussion). We then present, in FIGURE 1,

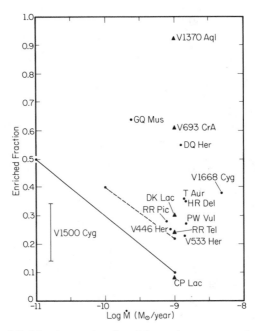

FIGURE 1. The enriched fraction as a function of the secular mean accretion rate (see text). In the systems denoted by *filled triangles,* the accretion rate cannot be determined (the orbital period is not known); these systems were placed arbitrarily at $\dot{M} = 10^{-9}$ $M_\odot$/yr. The *lines* represent the results of the diffusion calculations of Kovetz and Prialnik (9; *solid*) and (11; *dashed*). The range indicated for V1500 Cyg represents the difference between the result of Ferland and Shields[27] and that of Lance et al.[28]

the observed enrichments as a function of the accretion rates. The orbital periods of V1370 Aql, V693 CrA, DK Lac, RR Tel, and CP Lac are not known; these objects were arbitrarily placed at $\dot{M} \simeq 10^{-9}$ $M_\odot$/yr (which is comparable to most of the determined accretion rates). The accretion rate for V1500 Cyg was estimated on the basis of known constraints on its preoutburst brightness. Also shown in the figure are the results of Kovetz and Prialnik[9,11] (the points were arbitrarily connected by straight lines). Two features become immediately apparent from the figure: (1) *there appears to be no correlation between the observed enrichments and M̄, and (2) diffusion-induced convection generally predicts much lower enrichments than are observed.*

In view of the large uncertainties, both in the accretion rates and in the abundance determinations, no firm conclusion can be drawn from these apparent discrepancies. Taken at face value, however, the disagreement may mean either that the diffusion coefficients that have been used are incorrect (there are considerable uncertainties associated, for example, with the inclusion of screening effects), or that diffusion is not the dominant mechanism (at least in some cases) that produces the enrichments.

Another observational result that may be considered as a difficulty for the diffusion-induced convection model is the fact that the observed enrichments are highest for the two systems containing ONeMg white dwarfs (V1370 Aql and V693 CrA). These are expected to be very massive white dwarfs ($M_{\mathrm{WD}} \gtrsim 1.25 M_\odot$), and thus, unless the accretion rates in these systems happen to be unusually low, the outburst recurrence times for these systems should be relatively short ($\tau_{\mathrm{rec}} \propto R_{\mathrm{WD}}^4 / M_{\mathrm{WD}}$, see Truran and Livio[4]), implying relatively low enrichments via the diffusion mechanism.

Shear Mixing

The basic idea underlying the shear mixing mechanism is very simple. In nonmagnetic (or weakly magnetic) systems, the accretion disk extends all the way to the white dwarf surface. For material to actually be accreted, *at least a fraction of the angular momentum must be transferred to the white dwarf* (some angular momentum may be carried away by material flowing outward from the interaction region[18]). This leads on a short (dynamical) timescale to the formation of a boundary layer.[18–21] On the longer (accretion) timescale, this angular momentum must be transported to the interior, so that differential rotation is inevitably generated. *The presence of strong differential rotation is often associated with various hydrodynamical instabilities.* For example, the surface layers are subjected to the Kelvin–Helmholtz instability, expressed by the Richardson stability criterion

$$R_i \equiv \frac{N^2}{[r\,(\partial\Omega/\partial r)]^2} \geq \frac{1}{4}, \tag{1}$$

where N is the Brunt–Väisälä frequency (measuring the stabilizing effect of buoyancy) and Ω is the angular velocity. *Hydrodynamical instabilities can generate turbulence that, in turn, both redistributes angular momentum and causes elemental mixing.* Once small amounts of hydrogen are mixed into the core, ignition will occur, and mixing by convection can proceed in a similar manner to that associated with the diffusion mechanism discussed previously.

The general problem of shear mixing has been discussed extensively by Kippenhahn and Thomas,[22] MacDonald,[23] Livio and Truran[24] and, most recently, Fujimoto.[25] Fujimoto has shown that instabilities to nonaxisymmetric (adiabatic) perturbations and, in particular, the baroclinic instability, determine the distribution of angular momentum (except in the surface layers). He estimated that angular momentum will be transported inward, affecting a layer about 10–100 times thicker than the accreted layers. Elemental mixing, on the other hand, is less effective than angular momentum transport, due to the stabilizing effect of stratification (this is contrary to the simplifying assumption made by Kippenhahn and Thomas[22] that angular momentum and material mix in the same way). The ratio of (turbulent) material diffusivity v_m, to the turbulent (angular momentum transporting) viscosity, v_t, is roughly given by

$$\frac{v_m}{v_t} \simeq \frac{R_f}{R_i}. \tag{2}$$

Here R_f is the flux Richardson number that is of order 0.15 or lower,[26] and R_i is the Richardson number [Eq. (1)] that, in the layer in which the TNR occurs, can reach values of order

$$R_i \geq 400 \left(\frac{\Omega}{\Omega_k}\right)^2, \tag{3}$$

where Ω_k is the Keplerian angular velocity. Livio and Truran[24] pointed out the possible role of an Ekman-type spin-up process, in which one turbulent layer generates (by Ekman spin-up) a rotational shear layer, which then becomes turbulent, with the process repeating itself. It should be noted that estimates based on the baroclinic instability lead to dilation factors of the accreted material of the order of the observed ones under optimum conditions, $\Delta \equiv X_0/X \lesssim 5$, where X_0 and X are the hydrogen mass fractions in the accreted material and in the mixed layer.

The main points in favor of the shear mixing mechanism are the following:

1. In view of the large specific angular momentum of the material at the accretion disk–white dwarf boundary, it is difficult to see how at least some degree of shear mixing can be avoided. We should note perhaps that, if shear mixing operates, it tends to smooth out the composition discontinuity, thereby reducing the potential effects of diffusion.
2. The degree of enrichment is only weakly dependent on the accretion rate, a feature that is entirely consistent with the observations (FIG. 1).
3. The fact that the recurrent novae U Sco and V394 CrA do not show enrichments in heavy elements may also be consistent with the shear mixing model. In this case, this would be a consequence of the fact that, because recurrent novae accrete at a considerably higher rate than typical CNe (see Webbink et al.[12] for discussion), they accrete more angular momentum, which leads to higher values of Ω in the surface layers. This has the effect of strongly suppressing mixing. It can be shown that the dilation factor can be expressed as[25]

$$\Delta \simeq 1 + \frac{1}{2} R_f \left(\frac{\Omega_k}{\Omega}\right), \tag{4}$$

where it should be remembered that some experiments indicate that the Richardson flux number R_f decreases for large Richardson numbers R_i [see Eq. (3)]. Thus, the degree of mixing is reduced to very low values when Ω approaches Ω_k.

There are several difficulties currently associated with the shear mixing mechanism, of which we list two here:

1. To date, no self-consistent formalism for an actual calculation of the mixing process has been developed. This is, of course, a direct result of the fact that we are dealing with a three-dimensional, nonlinear process that involves hydrodynamical instabilities. A quantitative comparison with observations is therefore essentially impossible.

2. Shear mixing in its simple form (disk–white dwarf interaction) is not expected to operate for the strongly magnetized white dwarfs, since in those cases the disk is either disrupted or even nonexistent (the flow is dominated by the magnetic field). The AM Her object V1500 Cyg, however, shows a heavy element abundance[27,28] of $Z = 0.20 - 0.34$. While a different form of shear mixing may operate in these systems, when the material accumulated at the magnetic poles breaks the confinement and spreads over the entire surface,[29] this possible alternative mechanism has not been explored at all. Thus, the enrichments observed in V1500 Cyg may indicate that a different mixing mechanism is dominant (at least in magnetic systems).

We shall now discuss the mechanisms of the second type, for which mixing takes place during the TNR.

Convective-Overshoot-Induced Flame Propagation

The mechanism of mixing driven by convective overshoot was proposed by Woosley.[30] He used the convective-overshoot prescription employed by Weaver, Zimmerman, and Woosley,[31] in which nonconvective spatial zones that are immediately adjacent to convective regions are also slowly mixed, roughly on about ten times the radiative diffusion timescale (without energy transport). The actual diffusion coefficient that he employed was the minimum of (1) the value obtained from the convective velocity, when a 1 percent superadiabatic temperature gradient was assumed, and (2) 0.1 times the radiative diffusion coefficient.

When this procedure was employed *during the TNR,* when the hydrogen-rich shell had become fully convective, small amounts of hydrogen were found to be mixed into the CO core. Since the base of this mixed layer had a temperature comparable to that of the burning shell, every proton mixed inward was immediately captured. The associated release of energy led, eventually, to the linkage of the substrate by means of a diffusive wave into the white dwarf core. In the numerical study of Woosley,[30] it was found that $4 \times 10^{-6} M_\odot$ of carbon and oxygen were dredged up in this manner during the TNR (the accreted envelope was $\Delta m_{acc} \simeq 2.7 \times 10^{-5} M_\odot$).

The main successes of the convective-overshoot-induced flame-propagation model can be considered to be the following.

1. It relates the mixing process directly to the TNR, which is the cause of the explosion, and thus should certainly occur in all novae (including those possessing strong magnetic fields).

2. It does not appear to depend (certainly not strongly) on the accretion rate, in agreement with observations.

3. Its efficiency may be related to the efficiency of convection, which in turn may be related to the strength of the TNR.

This may explain why enrichments are higher for the more massive ONeMg white dwarfs, since the outbursts are stronger for such systems.

There are two main difficulties encountered by this model:

1. There are large uncertainties associated with the operation of the convective-overshoot mechanism. In particular, a two-dimensional hydrodynamical study of the ignition process, which did not assume mixing length convection, did not obtain any mixing between the hydrogen-rich layer and core material.[32] This was a direct consequence of the steep gradients at the core surface. Three-dimensional calculations will be required to settle this question.

2. The fact that the recurrent novae U Sco and V394 CrA do not show enrichments in heavy elements is inconsistent with the convective-overshoot mechanism. Since the mixing in this case is associated with the TNR itself, and is not a function of the details of the accretion process, the recurrent novae should experience comparable levels of enrichment. If future observations confirm that mixing does not occur in recurrent novae, this would therefore constitute a serious difficulty for the convection-overshoot-induced mixing model.

Convection-Induced Shear Mixing

Kutter and Sparks[33] have proposed a mechanism of convection-induced shear mixing, which effectively combines the action of shear mixing with the action of convection during the TNR. It is thus, in some sense, a combination of the shear mixing and convective-overshoot models. Kutter and Sparks suggested, specifically, the following sequence of events. The shear between the accretion disk and the white dwarf produces (by a Kelvin–Helmholtz instability) an accretion belt that rotates at near Keplerian velocities. In a particular calculation in which angular momentum was transported according to the Kippenhahn and Thomas[22] prescription (the generation of a layer that was marginally stable to the Richardson criterion), this situation was achieved after 7080 years. At this point shear mixing ceased (in the Kutter and Sparks formalism). Kutter and Sparks suggested that accretion continues (with no further mixing occurring) until the TNR. During the TNR, convection transports angular momentum from the outer (rapidly rotating) layers inward, causing the inner layers to rotate faster. This is presumed to lead to a new shear instability between the bottom layers of the convection zone and the white dwarf layers immediately below, resulting in mixing.

At present, the model of convectively induced shear mixing is too crudely formulated to allow a realistic evaluation of whether it is a viable mechanism. Its formulation ignores certain physical processes. For example, the configuration at which the initial mixing is assumed to stop (a rapidly rotating equatorial belt), is itself potentially unstable to various instabilities. In particular, the possible role of the baroclinic instability (described in our discussion of shear mixing) has been entirely ignored. In addition, the influence of hydrogen mixing into the core was not taken fully into account in the original formulation. Nevertheless, the idea that the convection that develops during the TNR may play an important role in the transport of angular momentum certainly deserves further study. We should point out, however, that because of the fact that the proposed mechanism of convection-induced shear mixing employs physical processes both from mechanisms that operate before the TNR (shear mixing) and from processes that operate during the TNR (convection), it can encounter difficulties associated with both types of mechanisms. For example, the convection-induced shear-mixing model will find both magnetic novae and recurrent novae difficult to explain (see our discussions of shear mixing and of convective-overshoot-induced flame propagation).

SUMMARY AND CRITICAL OBSERVATIONAL TESTS

Our survey of the different mixing mechanisms that have been proposed to explain the substantial levels of heavy-element abundances present in nova envelopes reveals that, at present, all the proposed mechanisms encounter some difficulties. Diffusion-induced convection and convective-overshoot-induced flame propagation are currently the only mechanisms for which there exist well-defined prescriptions for their calculation. Even in these cases, however, considerable uncertainty exists in the values of the relevant physical parameters (e.g., diffusion coefficients).

There exist several critical observations, which can possibly help in determining the relative importance of the different mechanisms.

1. Abundance determinations for the ejecta of those recurrent novae that are believed to be powered by thermonuclear runaways (in particular, T Pyx, which may erupt at any moment[12]). For these cases, it is extremely important to determine whether enrichments in heavy elements (or helium) do exist. If all such recurrent novae were found to be characterized by matter of solar composition, this would present serious difficulties for the convective-overshoot- and convection-induced shear-mixing models. Conversely, if the ejecta of the recurrent novae were found to be systematically enriched in heavy elements, it would deal a fatal blow to the diffusion model.
2. The identification of other magnetic (AM-Her-type) novae and abundance determinations in these objects. High enrichments in these objects, for which the disk should be disrupted, would present serious difficulties for the shear-mixing and convection-induced shear-mixing models.
3. Abundance determinations in "neon" novae. A consistently high enrichment level in these novae, which are believed to involve massive ONeMg white dwarfs, may again represent a major difficulty for diffusion-induced convection.

We must conclude that the mixing mechanism in classical novae has yet to be unambiguously identified. Abundance data provide an extremely useful tool toward obtaining an understanding of this mechanism and, thereby, an understanding of a large number of nonlinear processes associated with accretion and thermonuclear runaways in classical nova systems.

ACKNOWLEDGMENT

The authors wish to express their thanks to Professor R. Kippenhahn for the hospitality of the Max-Planck-Institut für Astrophysik, Garching bei München, during the summer of 1989.

REFERENCES

1. TRURAN, J. W. 1982. *In* Essays in Nuclear Astrophysics, C. A. Barnes, D. D. Clayton, and D. N. Schramm, Eds.: 467. Cambridge University Press. Cambridge, England.
2. TRURAN, J. W., J. HAYES, M. LIVIO & A. SHANKAR. 1990. Preprint.
3. TRURAN, J. W. 1990. *In* The Physics of Classical Novae. IAU Colloquium 122, A. Cassatella, Ed. In press.
4. TRURAN, J. W. & M. LIVIO. 1986. Astrophys. J. **308:** 721.
5. WILLIAMS, R. E. 1985. *In* Production and Distribution of CNO Elements, I. J. Danziger, Ed.: 225. ESO, Garching,
6. ANDERS, E. & M. EBIHARA. 1982. Geochim. Cosmochim. Acta **46:** 2363.
7. CAMERON, A. G. W. 1982. *In* Essays in Nuclear Astrophysics, C. A. Barnes, D. D. Clayton, and D. N. Schramm, Eds.: 23. Cambridge University Press. Cambridge, England.
8. PRIALNIK, D. & A. KOVETZ. 1984. Astrophys. J. **281:** 367.
9. KOVETZ, A. & D. PRIALNIK. 1985. Astrophys. J. **291:** 812.
10. MICHAUD, G., G. FONTAINE & Y. CHARLAND. 1984. Astrophys. J. **280:** 247.
11. KOVETZ, A. & D. PRIALNIK. 1990. *In* The Physics of Classical Novae. IAU Colloquium 122, A. Cassatella, Ed. In press.
12. WEBBINK, R. F., M. LIVIO, J. W. TRURAN & M. ORIO. 1987. Astrophys. J. **314:** 653.
13. MESTEL, L. & H. C. SPRUIT. 1987. Mon. Not. R. Astron. Soc. **226:** 57.
14. HAMEURY, J. M., A. L. KING, J. P. LASOTA & H. RITTER. 1988. Mon. Not. R. Astron. Soc. **231:** 535.
15. SHARA, M. M., M. LIVIO, A. F. J. MOFFAT & M. ORIO. 1986. Astrophys. J. **311:** 163.
16. LIVIO, M. 1990. *In* The Physics of Classical Novae. IAU Colloquium 122, A. Cassatella, Ed. In press.
17. PATTERSON, J. 1984. Astrophys. J. Suppl. Ser. **54:** 443.
18. KLEY, W. 1989. Astron. Astrophys. In press.
19. PRINGLE, J. E. & G. J. SAVONIZE. 1979. Mon. Not. R. Astron. Soc. **197:** 777.
20. PAPALOIZOU, J. C. B. & G. Q. G. STANLEY. 1986. Mon. Not. R. Astron. Soc. **220:** 593.
21. REGEV, O. & A. A. HONGERAT. 1988. Mon. Not. R. Astron. Soc. **232:** 81.
22. KIPPENHAHN, R. & H.-C. THOMAS. 1978. Astron. Astrophys. **63:** 625.
23. MACDONALD, J. 1983. Astrophys. J. **273:** 289.
24. LIVIO, M. & J. W. TRURAN. 1987. Astrophys. J. **318:** 316.
25. FUJIMOTO, M. Y. 1988. Astron. Astrophys. **198:** 163.
26. ROHR, J. J., E. C. ITSWEIRE & C. W. VAN ATTA. 1984. Geophys. Astrophys. Fluid Dyn. **29:** 221.
27. FERLAND, G. J. & G. A. SHIELDS. 1978. Astrophys. J. **226:** 172.
28. LANCE, C. M., M. L. MCCALL, & A. K. VOMOTO. 1988. Astrophys. J. Suppl. Ser. **66:** 151.
29. LIVIO, M., A. SHANKAR & J. W. TRURAN. 1988. Astrophys. J. **325:** 282.
30. WOOSLEY, S. E. 1986. *In* Nucleosynthesis and Chemical Evolution, B. Hauck, A. Maeder, and G. Magnet, Eds.: 1. Geneva Observatory. Savverng, Switzerland.
31. WEAVER, T. A., G. B. ZIMMERMAN & S. E. WOOSLEY. 1978. Astrophys. J. **225:** 1021.
32. OETTL, R. 1988. Ph.D. Thesis. Max Planck Institut für Astrophysik, Garching.
33. KUTTER, G. S. & W. M. SPARKS. 1989. Astrophys. J. **340:** 985.
34. WILLIAMS, R. E. & J. S. GALLAGHER. 1979. Astrophys. J. **228:** 482.
35. TYLENDA, R. 1978. Acta Astron. **28:** 333.
36. GALLAGHER, J. S., E. K. HEGE, D. A. KOPRIVA, R. E. WILLIAMS & H. R. BUTCHER. 1980. Astrophys. J. **237:** 55.
37. HASSAL, B. J. M., M. A. J. SNIJDERS, A. W. HARRIS, A. CASSATELLA, M. DENNEFELD, M.

FRIEDJUNG, M. BODE, D. WHITLET, P. WHITELOCK, J. MENZIES, T. L. EVANS & G. T. BATH. 1990. *In* The Physics of Classical Novae. IAU Colloquium 122, A. Cassatella, Ed. In press.

38. STICKLAND, D. J., C. J. PENN, M. J. SEATON, M. A. J. SNIJDERS & P. J. STOREY. 1981. Mon. Not. R. Astron. Soc. **197:** 107.

39. WILLIAMS, R. E., W. M. SPARKS, S. STARRFIELD, E. P. NEY, J. W. TRURAN & S. WYCKOFF. 1985. Mon. Not. R. Astron. Soc. **212:** 753.

40. WILLIAMS, R. E., N. J. WOOLF, E. K. HEGE, R. L. MOORE & D. A. KOPRIVA. 1978. Astrophys. J. **224:** 171.

41. SNIJDERS, M. A. J., T. J. BATT, P. F. ROCHE, M. J. SEATON, D. C. MORTON, T. A. T. SPOELSTRA & J. C. BLADES. 1987. Mon. Not. R. Astron. Soc. **228:** 329.

42. FERLAND, G. J. 1979. Astrophys. J. **231:** 781.

43. WHITLEY, B. A. & G. C. CLAYTON. 1989. Astron. J. **98:** 297.

Rapidly Rotating Neutron Stars: Implications of Half-Millisecond Periods[a]

JAMES R. IPSER

Department of Physics
University of Florida
Gainesville, Florida 32611

INTRODUCTION

The apparent observation[1] of a neutron star in SN1987A with a large rotational angular velocity $\Omega = \Omega_{SN} = 1.237 \times 10^4 \text{ s}^{-1}$ has focused attention on the constraints imposed on the high-density equation of state by such a large rotation rate. The purpose of this paper is to provide a discussion of these constraints and related implications for the origin, evolution, and stability of neutron stars. The foundation for the present discussion is the collection of relativistic neutron-star models that has been amassed by Friedman, Ipser, and Parker[2] through use of numerical codes that are adaptations of the numerical programs originally developed by Butterworth and Ipser.[3]

RESULTS

We shall summarize certain properties of uniformly rotating models based on a variety of equations of state. Many of these equations of state are in the collection of Arnett and Bowers,[4] and they are identified here by the symbols used in that reference. Additional equations of state that we employ are the Friedman and Pandharipande[5] equation of state (denoted FP) and the Weise and Brown[6] pion-condensate equation of state (denoted by π) with effective axial–vector coupling strength $g_A^* = 1.3$.

For each equation of state, a key model is the uniformly rotating model with the maximum gravitational mass. This model rotates with the maximum uniform angular velocity, for the given equation of state, that is consistent with the constraint of stability against gravitational collapse. (We ignore for the moment the question of nonaxisymmetric instabilities, which are not associated with gravitational collapse.) Equilibrium models whose rotation rates exceed that of the maximum-mass model do exist, but they are all unstable to gravitational collapse.[7]

Some important properties of the maximum-mass models for our collection of equations of state are exhibited in TABLE 1.

Comparison with nonrotating models reveals that Ω_{max}, the maximum allowed angular velocity for a given equation of state, satisfies the relation

$$\Omega_{max} \approx 0.72 \times 10^4 \text{ s}^{-1} \left(\frac{M_S}{M_\odot}\right)^{1/2} \left(\frac{R_S}{10 \text{ km}}\right)^{-3/2}$$

[a] This work was supported in part by National Science Foundation Grant PHY-8906915.

to within about 10 percent accuracy, where M_S and R_S are the mass and radius of the maximum-mass spherical model. That a relation of this form should exist was first suggested by Shapiro, Teukolsky, and Wasserman.[8]

The exhibited values for Ω_{max} and the observed value Ω_{SN} for SN1987A clearly place strong constraints on viable equations of state. Notice that the existence of uniformly rotating neutron stars with angular velocities $\Omega \geq \Omega_{SN}$ would conclusively rule out stiffer equations of state, including FP, C, D, and L. Of the exhibited equations of state that survive, F and A require a neutron star with $\Omega = \Omega_{SN}$ to be rotating at nearly the maximum allowed rate. One expects that a configuration this close to the limiting rate will exhibit unstable nonaxisymmetric modes of pulsation that radiate away angular momentum in gravitational waves, and thereby cause the configuration to spin down on a timescale $< 10^{11}$ s, the lower limit reported for SN1987A. In this connection, Ipser and Lindblom[9] have recently solved the problem of numerically constructing solutions to the normal-mode pulsation equations for

TABLE 1. Models with Maximum Mass and Rotation Rate for the Equations of State Mentioned in the Text

Equation of State	Ω_{max}	ϵ_c	$M/M_\odot$	(%)	$M_0/M_\odot$	R
L	0.76	1.11	3.18	(20)	3.72	17.30
D	1.04	2.78	1.94	(17)	2.21	12.70
C	1.11	2.71	2.16	(17)	2.47	12.90
FP	1.23	2.50	2.30	(17)	2.71	12.00
F	1.24	4.10	1.66	(13)	1.87	11.00
A	1.28	3.29	1.94	(17)	2.25	10.80
π	1.54	4.47	1.74	(15)	2.02	9.18
B	1.57	5.16	1.65	(17)	1.91	9.20
G	1.52	5.50	1.55	(14)	1.73	8.60

NOTE: The quantities listed are Ω_{max}, angular velocity in 10^4 s^{-1}; ϵ_c, central density in 10^{15} g cm^{-3}; $M/M_\odot$, gravitational mass; (%) percentage increase of gravitational mass above the maximum mass of spherical model; $M_0/M_\odot$, baryon mass; R, equatorial radius in kilometers.

rapidly rotating Newtonian configurations. Their results lead to the prediction that the critical angular velocity, Ω_c, at which nonaxisymmetric instability sets in is given by $\Omega_c \approx 0.9\Omega_{max}$ for a neutron star of the age of SN1987A. This prediction is based on the estimate that such a neutron star has cooled to temperature $T \sim 10^9$ K. This temperature and a viscosity model based on neutron–neutron nonsuperfluid viscosity have been used to estimate the stabilizing effects of viscous dissipation. It is important to note that if the temperature is actually significantly lower than the value just given, the viscosity could be strong enough to stabilize all nonaxisymmetric modes.

It thus appears that except for B, G, and π, all equations of state in our complex are incompatible with an observed angular velocity as large as Ω_{SN}.

In fact, equations of state as soft as B or G also appear to be ruled out, but for a different reason. Assuming the validity of general relativity, the measured mass of the binary pulsar is $(1.442 \pm 0.003)M_\odot$.[10] Since rotation has negligible effect on the mass of the binary pulsar, its measured mass rules out all equations of state, including B and G, that yield maximum nonrotating masses $\lesssim 1.442 M_\odot$.

The conclusion is that the only viable equations of state are rather soft ones that, like π, lie in a rather narrow range between B and A. If it turns out that the maximum nonrotating mass is significantly larger than the binary pulsar value—the value $(1.85 \pm 0.3) M_\odot$ has been reported for 4U0900-40[11]—the allowed range will become uncomfortably narrow.

For an allowed equation of state, a neutron star might evolve to $\Omega \sim \Omega_{SN}$ in the following way. Since viscosity $\propto T^{-2}$, and since Ω_c decreases with decreasing viscosity,[11] one pictures an initially hot ($T \gtrsim 10^{10}$ K), rapidly rotating state with $\Omega \gtrsim 0.9\Omega_{max} > \Omega_c$. In this state normal modes of pulsation with spherical harmonic indices $l = m \geq 3$ are unstable; and the neutron star spins down on a timescale[2]

$$\tau \sim 10^6 \left(\frac{M}{1.4 M_\odot}\right)^{-4} \left(\frac{R}{10 \text{ km}}\right)^5 \left[\frac{\Omega - \Omega_c}{0.1\Omega_c}\right]^{-(2m+1)} \text{ s,}$$

where R is the stellar radius. As the neutron star cools to $T \sim 10^9$ K, viscosity increases and Ω_c rises to $\sim 0.9\Omega_{max}$.[9] Hence Ω and Ω_c approach each other. In this phase, which would be the current one for SN1987A in the present picture, the spin-down timescale becomes[11,12]

$$\tau \sim 10^9 (\Omega - \Omega_c)^{-1} \left(\frac{M}{1.4 M_\odot}\right)^{1/2} \left(\frac{R}{10 \text{ km}}\right)^{1/2} \left(\frac{\epsilon}{10^{15} \text{ g cm}^{-3}}\right)^{-5/4} \left(\frac{T}{10^9 \text{ K}}\right)^2,$$

where ϵ is the mean density of the neutron star. Thus, SN1987A currently might be hovering near the instability point.

From our results, one can draw inferences for gravitational collapse and for the masses of neutron-star progenitors. For every equation of state in our complex, the baryon mass of a neutron star with $\Omega \geq \Omega_{SN}$ exceeds the maximum nonrotating baryon mass. If it spins down sufficiently far, such an object will become unstable to gravitational collapse.[7] Hence, one can speculate on the existence of a class of black holes formed in this way with masses $\sim 1.5 M_\odot - 2 M_\odot$. In addition, all our neutron-star models have gravitational masses $M \geq 1.5 M_\odot$ and baryon masses $M_0 \geq 1.7 M_\odot$ when $\Omega \geq \Omega_{SN}$. This places a lower limit $M \approx M_\odot \geq 1.7 M_\odot$ on the immediate progenitor of SN1987A (assuming negligible gravitational binding energy for the progenitor).

REFERENCES

1. KRISTIAN, J., C. R. PENNYPACKER, J. MIDDLEDITCH, M. A. HAMVY, J. N. IMAMURA, W. E. KUNKEL, R. LUCINIO, D. E. MORRIS, R. A. MULLER, S. PERLMUTTER, S. J. RAWLINGS, T. P. SASSEEN, I. K. SHELTON, T. Y. STEIMAN-CAMERON & I. R. TUOHY. 1989. Nature **338:** 234.
2. FRIEDMAN, J. L., J. R. IPSER & L. PARKER. 1986. Astrophys. J. **304:** 115.
3. BUTTERWORTH, E. M. & J. R. IPSER. 1976. Astrophys. J. **204:** 200.
4. ARNETT, W. D. & R. L. BOWERS. 1977. Astrophys. J. Suppl. Ser. **33:** 415.
5. FRIEDMAN, B. & V. R. PANDHARIPANDE. 1981. Nucl. Phys. **A361:** 502.
6. WEISE, W. & G. E. BROWN. 1975. Phys. Lett. **58B:** 300.

7. FRIEDMAN, J. L., J. R. IPSER & R. SORKIN. 1988. Astrophys. J. **235:** 722.
8. SHAPIRO, S. L., S. A. TEUKOLSKY & I. WASSERMAN. 1983. Astrophys. J. **272:** 702.
9. IPSER, J. R. & L. LINDBLOM. 1989. Phys. Rev. Lett. **62:** 2777.
10. TAYLOR, J. H. & J. M. WEISBERG. 1989. Princeton University Observatory Report. Princeton, N.J.
11. CUTLER, C. & L. LINDBLOM. 1987. Astrophys. J. **314:** 234.
12. LINDBLOM, L. 1986. Astrophys. J. **303:** 146.

Nonlinear Evolution of Protostellar Disks and Light Modulations in Young Stellar Objects[a]

D. N. C. LIN AND K. R. BELL

Lick Observatory
University of California Observatory
University of California
Santa Cruz, California 95064

INTRODUCTION

Excess infrared radiation detected in young stellar objects are often used as indicators of the presence of circumstellar disks.[1,2] Though unresolved directly, roughly half of all pre-main-sequence stars are deduced to have disks with masses in the range $0.01-0.1M_\odot$ and sizes from 10 to 100 AU.[3-5] In the case of HL Tau, where the image is resolved, a disk with a size 2000 AU and mass $0.1-1M_\odot$ is deduced from radial velocity maps.[6] Theoretical investigation of disk dynamics in this context is important for not only the analysis of these observed properties but also for the theory of formation of rotating and multiple stars.[7] Close comparison between the conditions that lead to the formation of the solar system and disks around young stellar objects can provide observable tests for theories of cosmogony.[8] At the same time, we can deduce signatures of protoplanetary formation on accretion disks around young stellar objects.

Accretion disk theory is based on the conjecture that the source of energy is viscous dissipation of differentially rotating gas in the disk. Associated with energy dissipation is angular momentum transfer and mass diffusion. The rate of disk evolution is determined by the magnitude of an effective viscosity.[10] Molecular viscosity is generally too small to make a significant contribution. Disk models provide estimates on the timescale and physical mechanism for the viscous evolution. Theoretical results can be tested by observations of nonsteady disk evolution. For example, if the infrared excess is due to viscous dissipation or reprocessing of stellar radiation, the disappearance of disks, as young stellar objects evolve toward a "naked" T Tauri phase, would imply a viscous evolution timescale $\sim 10^{5-7}$ years.[3]

Observation of accretion disks in steady state can provide useful information on the accretion rate.[1,11,12] Observational data and theoretical analyses of time-dependent disk flows can, however, provide more useful information that cannot be deduced from disks in a steady state. For example, from the evolutionary timescale,

[a]This work is supported in part by National Science Foundation Grants AST-86-21636 and AST-89-14173, and in part by NASA Grants NAGW 1211 and NGT50281. Part of this work has been conducted under the auspices of a special NASA astrophysics theory program that supports a Joint Center for Star Formation Studies at NASA–Ames Research Center, University of California, Berkeley, and University of California, Santa Cruz.

we can deduce the magnitude of viscosity. From the mass transfer rate and the magnitude of viscosity, the surface-density distribution can be estimated. Using the surface temperature and surface-density distribution, we can evaluate the optical depth and the midplane temperature.[8,13,14] These parameters are particularly important for constructing models of solar systems formation. In particular, the midplane temperature distribution is important for determining the chemical fractionation process. The surface density in the planet-forming region provides information on the availability of material out of which planets can be formed. From the magnitude of viscosity and the disk temperature, we can estimate the critical protogiant planet's mass for tidal truncation of the disk and termination of protoplanetary growth.[15]

Protostellar disks generally have finite sizes. At least in the outer regions of these disks, mass transfer rate cannot be independent of radius, and evolution is expected over the entire disk on a global viscous diffusion timescale that may be longer than 10^5 years. Nonsteady accretion flow is evident in many systems where the observed infrared continuum radiation does not follow that predicted by the steady-state accretion disk model.[1] Many young stellar objects such as FU Orionis stars and some T Tauri stars exhibit variations on the timescale of days to years.[16] These variations have been identified by modulations in mass transfer rate in the inner regions of protostellar disks.[17-19] In this case, unsteady flow is more likely to be caused either by intrinsic disk instabilities[8,20,21] or interaction with by external perturbers.[22]

In this paper, we present an evolutionary model for a protostellar disk. We first discuss various mechanisms for angular momentum transport in the next section, and the evolution of a viscous accretion disk in the third section. We present a general discussion on the evolution of eruptive phenomena in protostellar disks in the fourth section, and investigate a particular class of feedback mechanisms in the fifth section. In the last section, we discuss the implications of our results to solar-system formation.

EFFECTIVE TRANSPORT PROCESSES IN PROTOPLANETARY DISKS

Thermal Convection: A Mechanism for Heat and Angular Momentum Transport

In a differentially rotating geometrically thin disk, mass flow requires both angular momentum transport and energy dissipation.[10] These two processes can operate simultaneously through an effective viscous stress. Molecular viscosity in typical accretion disks is too small to be of astrophysical interest. In a variety of accretion disks, turbulent viscosity is often assumed to be responsible for both angular momentum transport and viscous dissipation despite the lack of rigorous proof that turbulence may occur intrinsically.[10,23] Protoplanetary disks, however, are unstable against thermal convection, in the direction normal to the plane of the disk, that is, the vertical direction, and convection drives the turbulent flow.[8,24]

In order to show that a protoplanetary disk is intrinsically unstable against thermal convection, we analyze the structure of a turbulent-free protoplanetary disk in which rotation prevents gas from migrating in the radial direction. However, gas can contract in the vertical direction. If the disk is not in hydrostatic equilibrium initially, it would rapidly evolve toward such a state. Using a one-dimensional numerical hydrodynamic scheme, Ruden[25] showed that a disk of cold gas contracts toward the midplane. During the initial rapid gravitational contraction phase, the released thermal energy can induce super adiabatic temperature gradients in the vertical direction to cause thermal convection.

After the disk has settled into a quasi-hydrostatic equilibrium, slow contraction continues as thermal energy is lost from the disk's surface. In the absence of any source of energy, surface radiation causes heat to diffuse from the midplane to the surface region. The associated reduction in the pressure support leads to a readjustment toward a new hydrostatic equilibrium. In the typical ambient temperature range of ten to a few thousand degrees, such adjustment leads to a super adiabatic structure in the vertical direction. This tendency is caused by the opacity, primarily due to dusts, being an increasing function of temperature. In this case, the cooler surface region has a lower opacity and cools more efficiently.[8,24,26] This condition is not generally satisfied in accretion disks as in other astrophysical contexts. According to the standard Schwarzschild convective stability criterion, a superadiabatic gradient induces the disk to become convectively unstable in the vertical direction.

The role of convection in a disk is not limited to heat transport in the vertical direction. Convective eddies can induce mixing over a radial extent comparable to their size, which ranges up to a fraction of the disk's thickness. Through this mixing process, angular momentum is transferred. Convection also generates turbulence that causes dissipation of energy stored in differential rotation. Perhaps the simplest treatment for convection is to use the mixing length prescription in which the eddy viscosity is assumed to be the product of convective speed and an effective mixing length that is comparable to the size of the eddies.[24] From such a treatment, we can build self-consistent models in which convection is responsible for (1) energy dissipation, (2) heat transport in the vertical direction, (3) angular momentum and mass transport in the radial direction.

The mixing-length model, though informative and easy to use, is based on an *ad hoc* prescription of eddy viscosity. In a convective disk, eddies with a variety of scales are generated. Similar to typical turbulent shear flows, the largest eddies often provide the dominant momentum transfer, whereas the smallest eddies provide most of the energy dissipation in the disk. The scale of the largest convective eddies is comparable to the entire convectively unstable zone, which itself is a significant fraction of the disk's thickness. On these large scales, global effects such as rotation and radiative losses are important. Thus, convection must be examined with a global analysis. In an attempt to carry out a global analysis, Cabot et al.[27,28] computed a vertically averaged effective viscosity that is derived from integrating the linear growth rate through various distances above the midplane of the disk. This growth rate varies greatly in the vertical direction, and therefore cannot be attributed to a given eddy.

A more appropriate global treatment is to determine a unique growth rate and its associated eigenfunction for each characteristic convective mode.[29] These eigenfunctions extend over finite radial distances. In the thin-disk limit, WKB approximation may be used to describe the radial dependence of temperature and density. These global linear stability analyses of axisymmetric perturbation indicate that (1) rotation and compressibility tend to reduce the growth rate of the disturbances; (2) the growth rate is proportional to the square root of the radial wave number and is bounded by the maximum values of the Brunt–Vaisala frequency; (3) the maximum radial size of eddies scales as the square root of superadiabaticity times the size of the convective region; (4) due to radiative losses, the short wavelength modes become overstable, and only the fundamental and the first harmonic modes, where the wavelength is comparable to thickness of the disk, can grow effectively; and (5) both even and odd modes exist in which a single eddy may either be confined to one side of or thread through the midplane and have a characteristic scale comparable to the thickness of the entire disk.

Convective eddies can provide a relatively effective coupling between different parts of the disk, as well as induce heat dissipation and angular momentum transport. The magnitude of the effective viscosity can be derived under the assumption that gas, within a radial wavelength, mixes efficiently over the characteristic growth timescale. This estimate generally agrees well with that derived on the basis of the mixing length model, that is, it yields an effective viscosity of the order $\alpha C_s^2/\Omega$,[23] where $\alpha \sim 0.01$ is dimensionless number, C_s and Ω are the sound speed and angular frequency in the disk. A more rigorous next step would be to carry out global analysis of the initial linear growth of nonaxisymmetric disturbances and its growth to nonlinear regions where dissipative processes become important. The determination of the torque associated with the growing nonaxisymmetric disturbance will provide a more rigorous estimate on the efficiency of angular momentum transport.

The superadiabatic gradient in the vertical direction can be significantly modified by the boundary condition at the surface of the disk. For example, during the formation stage of the disk, shock dissipation near the region where the infalling material joins the disk induces an isothermal structure for the disk. If the infall rate onto the disk remains constant, shock dissipation near the surface dominates energy loss due to contraction toward the midplane so that convection is suppressed. When the infall rate is reduced, however, super adiabatic gradients may be established as the surface heating becomes less important.[25]

Recently, Nagagawa, Watanabe, and Nakazawa[30] have shown that surface heating due to radiation from the central star can also reduce the temperature gradient in the vertical direction in a manner similar to shock dissipation associated with infall. In their calculation, Nagagawa *et al.* show that convection may be stabilized if there is a relatively large stellar radiation incident onto the surface of a disk with a relatively low surface density. Their choices of parameter are biased toward reducing the temperature gradient in the disk. A more general analysis[31] indicates that convection is suppressed only when the surface heating is sufficiently large to induce a blackbody temperature comparable to temperature near the midplane. This critical flux increases with the surface density of the disk, since opacity and consequently temperature near the midplane also increase with the surface density.

If the surface heating exceeds the critical flux required to stabilize against convection, it remains unclear how the disk may evolve subject to axisymmetric perturbation. Consider an axisymmetric perturbation in which one region of the disk has a slightly higher temperature. The disk thickness would increase there and expose that region of the disk to additional solar radiation. Although the disk may be stabilized against convection, this relatively thick region of the disk would cast shadows and reduce surface heating for the exterior regions. A decrease in the incident solar radiation in the shielded region would cause a reduction in the disk temperature. Consequently, the disk opacity is reduced and the cooling efficiency is increased. Although this process may not be thermally unstable, it may generate large temperature gradients in the radial direction across the interface between the exposed inner hot region and the shielded outer cool regions. For perturbations with wavelengths comparable to or shorter than the thickness of the disk, the disk may become convectively unstable in the radial direction. Mixing in the radial direction would limit the magnitude of radial temperature gradients. The consequence of surface heating may be the formation of ripples on the disk surface.

Nonaxisymmetric Gravitational Instabilities in the Outer Regions of Protostellar Disks

The discovery of massive disks around HL Tau[6] reveals the existence of disks whose self-gravity, in the vertical direction, is comparable to that due to the central star. The relative importance of self-gravity is measured in a dimensionless parameter Q.[32] When Q is of order unity, self-gravity of the disk may cause unstable growth to axisymmetric and nonaxisymmetric perturbations. In the context of angular momentum transfer, nonaxisymmetric instabilities are more interesting and relevant because not only do they induce nonaxisymmetric torque[33] but also, under certain circumstances, can exist even when the disk is stable against axisymmetric perturbations.[34,35]

The condition for gravitational instability against axisymmetric perturbations can be obtained through local analyses, and it is $Q < 1$.[32,36] The stability analyses of nonaxisymmetric perturbations generally require global analyses, because nonaxisymmetric perturbations induce torque that has strong effects over extended regions of the disk. Recently, Papaloizou and Lin[35] investigated the linear stability of self-gravitating rings and disks against nonaxisymmetric perturbations with global normal-mode analyses. These analyses showed that surface-density variation across corotation resonance is particularly important for inducing transmission of energy and angular momentum between the Lindblad and corotation resonances. For some surface-density variations, the disk becomes unstable against nonaxisymmetric perturbations, even when it is stable against axisymmetric perturbations, that is, $Q > 1$. The growth timescale for normal modes with Q slightly greater than unity and large azimuthal wave number, m, is several orbital periods at the corotation radius. The disk is stabilized when Q is substantially greater than unity. When the disk is very massive and has a relatively sharp edge, global $m = 1$ mode may also be excited.[7,37] The global $m = 1$ modes may be relevant for the formation of binary stars.

These results agree well with numerical simulations.[34] In these numerical simulations, the initial stellar velocity dispersion is sufficiently large that $Q = 1.5$ through the disk and no axisymmetric perturbation can grow. However, nonaxisymmetric perturbations grow exponentially. The growth occurs first near the inner boundary region where the growth timescale is minimum. The perturbations become saturated and the velocity dispersion increases so that the disk is stabilized. This secular evolution introduces variations in the effective surface density, which promotes the excitation and growth of nonaxisymmetric perturbations at a new location with some larger disk radius. Consequently, the fastest growing unstable region migrates outward until the entire disk is heated by the saturated growth of nonaxisymmetric perturbations. Thereafter, the effective value of Q is substantially increased. Unless there is dissipation of dispersive motion, the disk becomes stabilized.

Based on these analyses, it is tempting to derive an approximate formula for the effective torque generated by nonaxisymmetric instabilities in self-gravitating disks.[38] This approximation formula, though very crude, is easily applied to the analyses of the global evolution of a self-gravitating disk.[39] Note that the effective torque induced by gravitational instability is due to global transfer, and it would not necessarily lead to a turbulent flow pattern.

Wave Propagation as a Carrier of Energy and Angular Momentum

An alternative scheme for angular-momentum transport has been proposed by Shu,[40] Donner,[41] and Spruit.[42] In this scheme, self-similar shock waves may induce an effective angular-momentum transport in accretion disks. The self-similarity of these shock waves implies that the shock waves have zero pattern speed, and are therefore stationary. Wavelike disturbances induced by perturbation at very large disk radii may steepen into shock waves as they propagate inward. This requires that there be little or no dissipative damping of wave action as the waves propagate inwards. If these shock waves can be induced, the radial spacing between successive shock fronts continues to reduce. When the spacing between shock fronts is reduced to less than the vertical scale height of the disk, wave propagation in the vertical direction becomes important.

Although nonlinear wave propagation in a disk remains to be analyzed, a three-dimensional linear analysis has been carried out.[43,44] Due to differential rotation, the wavelength of linear waves is reduced as they propagate through the disk. When the wavelength becomes comparable to the vertical scale height, wave propagation becomes influenced by the vertical structure of the disk. In the optically thick region of a protoplanetary disk, the vertical structure is thermally stratified such that the temperature and sound speed decreases with distance from the midplane. The magnitude of the pressure gradient in the vertical direction is also larger than that in the radial direction. Consequently, a wavefront can become retarded as a wave propagates from the regions where it is launched. The refraction effect promotes upward propagation such that waves are confined in a limited radial extent. For moderate temperature contrast between the disk's midplane and its surface, waves with wavelength comparable to or shorter than the vertical scale height, can be transmitted into and dissipated at the tenuous upper atmosphere. Thus, wave propagation is not an effective mechanism for angular-momentum transport over extended regions in the disk. Nevertheless, on a local scale, wave propagation can lead to effective dissipation. If the energy in the shear can be continually and effectively transferred into waves through some unstable growing modes, wave energy dissipation may provide a significant source of local viscosity.

These results indicate that dissipative damping may prevent linear waves from steepening into shock waves. Even if shock waves are induced, refraction effects may cause the shock waves to bend and propagate in the vertical direction. Thus, we remain pessimistic that self-similar propagation of shock waves can be an effective mechanism for angular-momentum transport. It is of interest to consider the dissipation of the wave as it propagates up into the optically thin regions of the disk atmosphere. In a recent investigation, Murray and Lin[45] showed that, in the inner region of the disk near the protostar, shock dissipation of waves can trigger thermal instability in the disk atmosphere. If the accretion rate onto a protostar is sufficiently large, the energy dissipated above the disk atmosphere may be sufficient to generate an ionized wind.

DYNAMICAL EVOLUTION

The evolution of protostellar disks can be analyzed with a time-dependent diffusion equation[8,10]

$$\frac{\partial \Sigma}{\partial t} - \frac{3}{r} \frac{\partial}{\partial r} \left(r^{1/2} \frac{\partial}{\partial r} (\Sigma \nu r^{1/2}) - \frac{2S_\Sigma(r, t) J(r, tg)}{\Omega} \right) - S_\Sigma(r, t) = 0, \tag{1}$$

where Ω, Σ, and ν are the angular frequency, surface density, and viscosity of the disk material, respectively, $S_\Sigma(r,t)$ and $J(r,t)$ are the mass flux and excess angular momentum of the infalling material, respectively. Clearly, disk evolution depends on both the magnitude of viscosity and the properties of the infalling material.

In the investigation of a protostellar disk, the epoch of disk formation is an important stage. During the initial formation of the disk, the specific angular momentum carried by the infalling material may be particularly important in determining the structure of the disk. Since the mass of the central star is relatively small, the effect of self-gravity may be important. Nonaxisymmetric gravitational instability, if and when it occurs, generates torque that induces energy and angular momentum transport in the disk. Both of these effects can be incorporated into the treatment of the diffusion equation.[39] If the growth timescale for nonaxisymmetric perturbations is considerably longer than the local dynamical timescale, a substantially massive disk may be formed and preserved for a prolonged period of time. Disk masses for young T Tauri stars are estimated to be $\sim 0.01-1 M_\odot$.[4-6] Massive disks may also be needed to provide an environment for massive outbursts. Finally, massive disks may be needed to form binary stars.[7] Since binaries are very abundant, the occurrence of massive disks may also be frequent.

In the postinfall stage, viscous diffusion dominates the evolution of the disk. It is of interest to consider the evolution of a "minimum mass" solar nebula out of which the solar system was formed.[46,47] In this model, surface density distribution is derived by augmenting gas to the present mass distribution in the solar system based on the assumption that the protoplanetary disk had a solar composition and planetary formation is totally efficient at retaining all the heavy elements in the disk. Applying the effective viscosity associated with convective viscosity into the diffusion equation, we find the evolution of the minimum-mass nebula with a physical dimension comparable to that of the present-day solar nebula evolving on the timescale of $\sim 10^6$ years.[8] The temperature distribution resembles that deduced from the condensation temperature for various terrestrial planets and satellites.[48] At radii interior to Mercury, the midplane temperature exceeds 2000 K such that grains would be mostly evaporated.

For the minimum-mass solar-nebula model, the surface density of the disk is so low that self-gravity becomes at best marginally important beyond the orbit of Neptune, even when the disk temperature is 10 K. If the outer region is both optically thin and non-self-gravitating, the disk terminates its evolution there. This may be the reason why there is no major planet beyond the orbit of Neptune. Inside the outermost region, the disk remains opaque and viscous evolution continues. A fraction of disk material would be deposited into the outermost region to carry the excess angular momentum. While gas may eventually be evaporated by the photodissociation process, dust particles would be left behind. In the absence of turbulence, dust particles descend toward the midplane. When the dust layer becomes sufficiently thin, gravitational instability would cause the dust to clump and form 10-km-size objects that may be the pregenitors of planetesimals or comets.[49]

On the evolutionary timescale of typical T Tauri stars, $\sim 10^6$ years, a significant fraction of these young stellar objects become naked T Tauri, that is, they lose any indication of circumstellar disk. If the convectively driven turbulent viscosity were applied to Eq. (1), we deduce the timescale for the disk to evolve to an optically thin system to be $\sim 10^{8-9}$ years. Even assuming that some hypothetical transonic turbulence may be responsible for angular-momentum transfer, the timescale for the disk to evolve into an optically thin state is still larger than 10^7 years. These arguments imply that the disappearance of IR and UV excess is not entirely due to viscous diffusion.

One possible mechanism for eliminating the UV and IR excess from the disk is through dust settling. For example, surface heating effects can stabilize against convection, provided the surface density of the disk is sufficiently low.[30,31] During the evolution of the disk, the surface density of the disk continually decreases. For a minimum-mass nebula model, the surface density of the outer regions of the disk is reduced to a sufficiently low value, after a few million years, such that the surface heating effect stabilizes these regions against thermal convection. When the disk is stabilized, turbulence will decay unless there are other instabilities. These processes would cause the protoplanetary disk to become transparent so that it can no longer reprocess radiation from the central star. Furthermore, the lack of turbulent viscosity implies that there would not be any viscous dissipation to heat the disk or to supply disk mass to be accreted by the central star. Through this mechanism, radiative flux from the disk is significantly reduced, while most of the disk gas is retained. The disk gas may be eventually eliminated on a somewhat longer timescale by (1) stellar wind ablation, (2) wind caused by the dissociation or ionization of disk gas by the incident solar radiation, and (3) disk–protoplanet tidal interaction.

It is also possible that convective instability is not the only source of turbulence. In this case, although the surface heating effect may induce an isothermal vertical structure and eliminate convection, it also provides a more favorable condition for sound waves to propagate a large distance. Note that in the absence of vertical thermal stratification, the refraction of waves is eliminated. Consequently, when the surface density has been reduced below the critical value, the efficiency of angular-momentum transport may actually increase despite the lack of convection.

FU ORIONIS OUTBURST PHENOMENA AND UNSTEADY ACCRETION DISK FLOWS

Many protostellar objects undergo eruptive variations. The best examples are the FU Orionis stars, which are very luminous young stellar objects with considerable infrared excess. These systems undergo large outbursts with a rise in timescale of the order a few months to years.[16] A large fraction of the energy released during FU Orionis outbursts originates from the disk, since the broad infrared excess requires a temperature distribution rather than a single black body temperature.[17-19] The observed rotational broadening of absorption lines decreasing with wavelength is consistent with the expectation that the long wavelength radiation is produced by the outer regions of the disk.[11,50]

An hypothesis for unsteady mass transfer in a disk is based on the preoutburst spectrum of V1057 Cyg being that of a typical T Tau star.[16] Accretion rate in T Tauri stars is observed to be 10^{-8} to $10^{-7} M_\odot$ y^{-1},[12] whereas that for FU Orionis during an outburst is $\sim 10^{-4}$ to $10^{-3} M_\odot$ y^{-1}. The rise timescale for FU Orionis events is characteristically a few months to years. This is comparable to the dynamical timescale at a few AUs. Such a relatively short rise-timescale implies that variations in accretion occurred in the inner regions of the disk close to the accreting star. If the evolution in the disk is determined by viscous processes and the rise proceeds on a viscous diffusion timescale, the region of sudden increase in accretion rate would be confined to within 10^{12} cm, which is only slightly larger than the dimension of the accreting protostar. During the rise, however, the inferred disk mass being accreted by the star is $\sim 10^{-3} M_\odot$. If all of the accreted material originated from such a narrowly confined region, the surface density of the disk would be 10^{5-6} g cm^{-2}, which is about one thousand times larger than that inferred from the "minimum-mass nebula" model. It is of interest to note that the minimum-mass nebula model implies that a mass of $10^{-3} M_\odot$ is distributed over $\sim 10^{14}$ cm.

Modulations in the mass transfer rate may arise from more extended regions, provided the disk can respond on a timescale shorter than that associated with viscous diffusion. For example, in dwarf novas, where outbursts with up to five magnitude increases have been observed, the rise timescale, which is considerably shorter than the viscous-diffusion timescale, is attributed to the rapid propagation of a transition front in a thermally unstable disk.[51] The thermo- and hydrodynamic properties of the inner region of a protoplanetary disk, with a mass transfer rate $\sim 10^{-5} M_\odot$ y^{-1}, are somewhat similar to those throughout dwarf nova disks,[8] that is, in both cases, gas is partially ionized and the opacity is a rapidly increasing function of temperature. These regions are subject to a thermal instability if the magnitude of viscosity is an increasing function of temperature.[52-55] Theoretical disk evolution calculations[8,20,21] indicate that protostellar disks can indeed become thermally unstable if the surface density exceeds certain critical values that are functions of radius. At a fraction of AU, the critical surface density is $\sim 10^4$ g cm^{-2}, which is about an order of magnitude larger than that according to the minimum-mass nebula model.

Unlike dwarf novas, where the outbursts are sustained for only a few days, FU Orionis has remained in a postoutburst high state for more than 30 years. A similar slow decline is also observed in V1057 Cyg. Theoretical models[20] indicate that after an upward transition, the thickness becomes a significant fraction of the radius such that advective transport of heat stabilizes against thermal runaway in the disk region exterior to the thermal transition front. Consequently, propagation of the transition front may be prevented. This stabilizing effect is only effective for disks with a mass transfer rate comparable to or larger than a few times $10^{-5} M_\odot$ y^{-1}.

NONLINEAR FEEDBACK, A MECHANISM FOR MASS TRANSFER MODULATION

While outbursts in dwarf novas may be regulated by thermal instability alone,[56] the observed large magnitude and long duration indicate that additional large-amplitude perturbations or other instabilities may be needed to trigger and sustain FU Orionis outbursts.[21] There are several possible mechanisms that may induce large-amplitude perturbations to trigger thermal instabilities in the disk. For example, the passage of hypothetical companions through the inner regions of the disk may cause significant perturbations to the surface density and mass transfer rate in the disk.[22] In this scenario, the perigee passage of the companion needs to come within ~ 1 AU, and the mass of the perturber needs to be a significant fraction of that of the primary. Observational tests for this scenario are feasible. For example, if the companion's mass is sufficiently large, it may also accrete gas from the disk. For FU Orionis, the hypothetical companion must have traveled to several tens AUs, since the perigee passage that triggered the initial outbursts. At the distance of Orion, the expected angular separation between the primary and the hypothetical companion may be so sufficiently large that they are resolvable by a speckle interferometer.

Another class of possible models is through feedback limit cycles. A feedback mechanism may be induced by the dissipated energy released in the disk when the accretion rate is high. For example, the disk's surface may be exposed to energy generated in the inner regions of the disk. When the mass accretion rate is sufficiently large, the surface radiation may significantly modify the structure and transport properties of the disk. There are two possible scenarios through which the mass transfer process may be modified. In most intermediate regions, where the disk radius is about a few AU, the dominant source of effective viscosity may be intrinsic turbulence induced by convective instability.[24,26,29] If convectively driven turbulence is the only source of effective viscosity, accretion flow through the disk could be quenched by surface heating, since it would stabilize against convection by inducing an isothermal vertical structure.[30,31]

Alternatively, the inferred disk in FU Orionis systems is sufficiently large such that the outer regions of the disk may be marginally gravitationally unstable just prior to the outburst. Growth of nonaxisymmetric perturbations can lead to gravitational torque that induces mass- and angular-momentum transport.[7,35,38] In this case, large surface heating can also stabilize the disk and thereby reduce the effective mass transfer rate through the disk. Gravitational instability is more easily realized in the extended outer regions of the disk. However, the evolution timescale for these regions is relatively long. Although gravitational instability may play a role in regulating FU Orionis outbursts, it is unlikely to be associated with light variations in T Tauri stars.

In either case, a potential self-regulated feedback mechanism may be induced if the surface heating is primarily due to the energy dissipation rate at the inner region of the disk. In particular, a large mass transfer rate would induce a high rate of dissipation near the inner region of the disk. The enhanced flux would reduce the effective viscosity and thereby quench the mass diffusion through the disk. A reduction in the mass flux through the disk would lead to a reduction in the energy dissipation rate near the disk's inner region such that relatively large effective viscosity may be restored. A recent theoretical model of such a feedback process[31] showed that depending on the response of the disk-to-surface heating, the feedback process may lead to both regular limit-cycle modulation or chaotic evolution.

In order to study the generic evolutionary pattern, we introduce an *ad hoc* viscosity prescription in which we assume the surface heating effect is directly determined by the accretion rate at the disk's inner boundary, $\dot{M}_*(t)$. The viscosity prescription we adopt is

$$\nu = \nu_0 \left(\frac{1 + A(r)}{1 + A(r)(\dot{M}_*(t)/\dot{M}_a)^n} \right), \qquad (2)$$

where $\dot{M}_a$ and ν_0 are constants, and the position function, $A(r) = [4(r/R)^4(1 - (r/R)^4)]^{10}$, was introduced to confine the disk response to a limited radial extent. We have tried other position functions, and the qualitative properties of our results do not depend on the form of $A(r)$. At the outer boundary, R, a steady infall flux, $\dot{M}_0$, is introduced. The sensitivity index, n, is determined by the response of the disk to variations in $\dot{M}_*$. The values of $\dot{M}_0/\dot{M}_a$ and n are the model parameter that determines the evolution of the disk.

We adopted the preceding form for computational convenience. It has the basic property that $\nu \sim \nu_0$ for $\dot{M}_* < \dot{M}_a$, and ν monotonically vanishes for large $\dot{M}_*$. We have adopted different prescriptions for viscosity with feedback. The results presented below do not depend on the actual form of these prescriptions. For the present prescription, when $\dot{M}_0 < \dot{M}_a$, the flow is stable for all values of n. For $\dot{M}_0 = \dot{M}_a$, the flow is stable for small values of n and oscillatory for $n > 20$. For larger values of n, the oscillations are steady but of small amplitude (~ 15 percent). For larger values of n, the oscillation stays the same, but their frequency increases slowly with n. When $\dot{M}_0 > \dot{M}_a$, the flow becomes oscillatory at relatively low values of n. As n increases, period doubling occurs and eventually for sufficiently large n, disk evolution becomes chaotic.

As an example of the evolution of the disk from a stable system (constant mass flux) through a periodic oscillating system to a chaotically oscillating system, we present here a bifurcation sequence observed for a constant $n = 20$ with $\dot{M}_0/\dot{M}_a$ (hereafter defined as parameter B) slowly increasing from 1 to 10. For B only slightly greater than 1 ($B = 1.1$), the system pumps itself into a regular oscillating state. The feedback is successful in this because of the existence of a finite diffusion timescale, that is to say, there is a finite delay between a mass flux perturbation in the active region of the disk: $M(R_{active})$, and the resultant change in mass flux at the central object: $\dot{M}_*$. In this stable oscillating system, just as a mass flux enhancement wave reaches the innermost part of the disk, the luminosity of the central object is recovering from the previous enhancement wave. Because of the diffusion and inherent friction in the disk system, the oscillations reach some finite amplitude and do not cause a global instability.

As the parameter B is increased still further, the disk gets into a situation in which mass flux enhancement waves are reaching the central object before the effects of the previous wave have completely dissipated. This results in a slightly larger luminosity enhancement than in the previous situation, but it is followed by a slightly smaller luminosity rebound (i.e., the next mass flux peak will be depressed with respect to the first) because so much of the mass was drained onto the star in the first outburst. Note that the time averaged ratio of $\dot{M}_*/\dot{M}_a$ is equal to B, which is to say that the time averaged $\dot{M}_*$ is equal to $\dot{M}_0$.

In FIGURES 1–4, we present four examples of oscillations observed in the bifurcation sequence as previously outlined. In the first frame, we see that the mass flux through the central-most zone (and therefore the luminosity of the central object) oscillates steadily around the value of B as defined in the caption. In the second frame with a slightly larger mass input rate, we see that the system alternates steadily between two different maximums, and in the third, between four. In the fourth frame, the system has become fully chaotic. Note that the largest peaks appear to occur almost regularly (with some notable exceptions). With each form of viscous

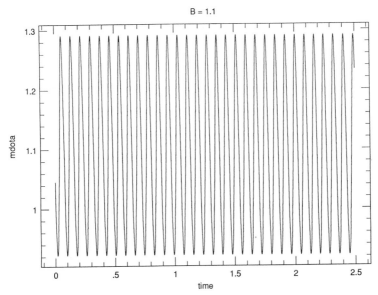

FIGURE 1. Disk evolution subject to feedbacks with $B = 1.1$. Single periodic oscillation is induced.

feedback and with each assumption about the unperturbed viscosity, ν_0, which were tested (details in Bell, Lin, and Ruden[31]) this general trend holds true.

We can provide some observational constraints on the feedback scenario. For example, if surface heating is important in quenching mass- and angular-momentum transfer, surface heating would become increasingly evident during the decline from the outbursts. Surface heating effects may be most easily detectable in the far-infrared continuum.[11] In addition, surface heating would induce an isothermal structure for disk atmosphere such that the absorption lines, if they exist at all, would be very shallow. The surface heating effect is expected to reduce to a minimum just prior to each outburst. If the restoration of relatively high effective viscosity is initiated at a relatively large disk radius, we would expect there to be some

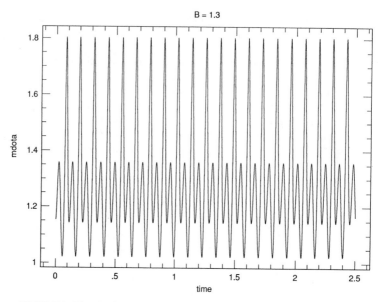

FIGURE 2. Biperiodic oscillations are induced by the feedback with $B = 1.3$.

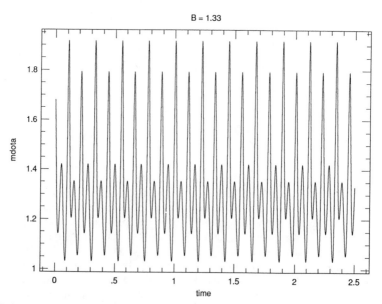

FIGURE 3. Fourfold periodic oscillations are induced by the feedbacks with $B = 1.33$.

far-infrared precursors to each outburst as a thermal transition front is evolved inwards. Theoretical models indicate that when the feedback process becomes nonlinear, the short timescale behavior can be very chaotic and irregular. But relative large-amplitude outbursts retain some regular pattern because they evolve on the viscous diffusion timescale of the maximum radial extent over which the disk participates in the outburst and feedback. We intend to construct models of spectral evolution and compare them with observational data to determine the radial extent of the outburst region. These data can also provide indication on the surface density distribution and viscosity in the inner regions of the disk. Thus, these data are most useful and important for constructing quantitative models for the origin of solar systems.

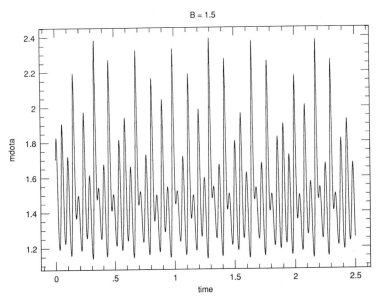

FIGURE 4. Chaotic oscillations are induced with $B = 1.5$.

DISCUSSION

In this paper, we reviewed some important aspects of various dynamical processes in the protoplanetary disk. Mass redistribution is determined by angular-momentum transfer. The rate of angular-momentum transfer is regulated by the magnitude of the effective viscosity. During the formation of the protoplanetary disks, mixing of infalling material with the disk gas can lead to significant mass transfer. Rapidly rotating clouds can lead to the formation of extended disks. In the outer regions of the disk, self-gravity of the disk can promote growth of nonaxisymmetric disturbance. The tidal torque associated with these growing unstable modes induces angular-momentum transport and regulates mass transfer. During the main phase of disk evolution, convectively driven turbulence provides an effective viscosity such that the typical evolutionary timescale of the protoplanetary disk is of the order $\sim 10^6$ years. This timescale is consistent with both the typical age of young stellar objects with signatures of circumstellar disks. It is also consistent with the timescale derived from the condition for tidal truncation of protoplanetary disks by proto Jupiter. If convective turbulence can be suppressed by radiative heating for even a small period of time, dust, the principal source of opacity in such systems, could settle out onto a flattened disk, perhaps preventing the recurrence of convection, and leading eventually to the break up of the disk and to the formation of planets. Understanding local and global influences on the effects of viscosity that exists in protostellar disks may thus be the key to understanding solar systems such as our own.

ACKNOWLEDGMENT

Aspects of the work reported here are due to P. H. Bodenheimer, C. Clarke, L. Hartmann, J. C. B. Papaloizou, J. E. Pringle, S. Ruden, G. Savonije, J. A. Sellwood, and F. Shu, whose contributions are greatly appreciated.

REFERENCES

1. ADAMS, F. C., C. LADA & F. H. SHU. 1987. Astrophys. J. **317:** 788.
2. SHU, F. H., F. C. ADAMS & S. LIZANO. 1987. Ann. Rev. Astron. Astrophys. **25:** 23.
3. STROM, S. E., S. EDWARDS & K. M. STROM. 1989. *In* Formation and Evolution of Planetary Systems, H. A. Weaver and L. Danly, Eds.: 91. Cambridge University Press. Cambridge, England.
4. ADAMS, F. C., J. P. EMERSON & G. A. FULLER. 1990. Astrophys. J. Preprint.
5. HARTMANN, L. & S. J. KENYON. 1990. Preprint.
6. SARGENT, A. I. & S. BECKWITH. 1987. Astrophys. J. **323:** 294.
7. ADAMS, F. C., S. P. RUDEN & F. H. SHU. 1988. Astrophys. J. **326:** 865.
8. LIN, D. N. C. & J. PAPALOIZOU. 1985. *In* Protostars and Planets II, D. Black and M. S. Matthews, Eds.: 981. University of Arizona Press. Tucson.
9. LIN, D. N. C. 1989. *In* Theory of Accretion Disks, F. Meyer, W. J. Duschl, J. Frank, E. Meyer-Hofmeister, Eds.: 89. Kluwer. Dordrecht, The Netherlands.
10. LYNDEN-BELL, D. & J. E. PRINGLE. 1974. Mon. Nut. R. Astron. Soc. **168:** 603.
11. KENYON, S. J. & L. HARTMANN. 1987. Astrophys. J. **323:** 714.
12. BERTOUT, C., G. BASRI & J. BOUVIER. 1988. Astrophys. J. **330:** 350.
13. LIN, D. N. C. & P. BODENHEIMER. 1982. Astrophys. J. **262:** 768.
14. RUDEN, S. P. & D. N. C. LIN. 1986. Astrophys. J. **308:** 883.
15. LIN, D. N. C. & J. PAPALOIZOU. 1986. Astrophys. J. **307:** 395.
16. HERBIG, G. H. 1977. Astrophys. J. **217:** 693.
17. HARTMANN, L. & S. KENYON. 1985. Astrophys. J. **299:** 462.
18. ————. 1987. Astrophys. J. **312:** 243.
19. ————. 1987. Astrophys. J. **322:** 393.
20. CLARKE, C., D. N. C. LIN & J. PAPALOIZOU. 1989. Mon. Not. R. Astron. Soc. **236:** 495.
21. CLARKE, C., D. N. C. LIN & J. E. PRINGLE. 1990. Mon. Not. R. Astron. Soc. **242:** 439.
22. CLARKE, C. & J. E. PRINGLE. 1990. In preparation.
23. SHAKURA, N. I. & R. A. SUNYAEV. 1973. Astron. Astrophys. **24:** 337.
24. LIN, D. N. C. & J. C. B. PAPALOIZOU. 1980. Mon. Not. R. Astron. Soc. **191:** 37.
25. RUDEN, S. P. 1986. Ph.D. Thesis, University of California, Santa Cruz.
26. LIN, D. N. C. 1981. Astrophys. J. **242:** 780.

27. CABOT, W., V. M. CANUTO, O. HUBICKYJ & J. B. POLLACK. 1987. Icarus **69:** 387.
28. ———. 1987. Icarus **69:** 423.
29. RUDEN, S. P., J. PAPALOIZOU & D. N. C. LIN. 1988. Astrophys. J. **329:** 739.
30. NAKAGAWA, Y., S. WATANABE & K. NAKAZAWA. 1988. *In* The Formation and Evolution of Planetary Systems, H. A. Weaver, F. Parsce, and L. Danly, Eds. Cambridge University Press. Cambridge, England. In press.
31. BELL, K. R., D. N. C. LIN & S. P. RUDEN. 1990. Astrophys. J. In press.
32. TOOMRE, A. 1964. Astrophys. J. **139:** 1217.
33. PAPALOIZOU, J. & G. SAVONIJE. 1990. Mon. Not. R. Astron. Soc. Preprint.
34. SELLWOOD, J. A. & D. N. C. LIN. 1989. Mon. Not. R. Astron. Soc. **240:** 911.
35. PAPALOIZOU, J. C. B. & D. N. C. LIN. 1989. Astrophys. J. **344:** 645.
36. SAFRONOV, V. S. 1960. Ann. d'Astrophys. **23:** 901.
37. SHU, F. H., S. TREMAINE, F. C. ADAMS & S. RUDEN. 1990. Astrophys. J. Preprint.
38. LIN, D. N. C. & J. E. PRINGLE. 1987. Astrophys. J. **225:** 607.
39. ———. 1990. Astrophys. J. **358:** 515.
40. SHU, F. H. 1976. *In* Structure and Evolution of Interacting Binary Stars, P. P. Eggleton, S. Mitton, and J. A. J. Whelan, Eds. Reidel. Dordrecht, The Netherlands.
41. DONNER, K. J. 1979. Ph.D. Thesis, Cambridge University, Cambridge, England.
42. SPRUIT, H. C. 1987. Astron. Astrophys. **184:** 173.
43. LIN, D. N. C., J. PAPALOIZOU & G. SAVONIJE. 1990. Astrophys. J. **364:** 326.
44. ———. 1990. Astrophys. J. In press.
45. MURRAY, S. & D. N. C. LIN. 1991. Astrophys. J. In press.
46. CAMERON, A. G. W. 1978. Moon Planet **18:** 5.
47. HAYASHI, C. 1981. Prog. Theor. Phys. Suppl. **70:** 35.
48. LEWIS, J. S. 1972. Icarus **16:** 241.
49. GOLDREICH, P. & W. R. WARD. 1973. Astrophys. J. **183:** 1051.
50. GRASDALEN, G. L., G. SLOAN, N. STOUT, S. E. STROM & A. D. WELTY. 1989. Astrophys. J., Lett. **339:** L37.
51. LIN, D. N. C., J. C. B. PAPALOIZOU & J. FAULKNER. 1985. Mon. Not. R. Astron. Soc. **212:** 105.
52. MEYER, F. & E. MEYER-HOFMEISTER. 1981. Astron. Astrophys. **104:** L10.
53. SMAK, J. 1982. Acta Astron. **32:** 199.
54. CANNIZZO, J. K., P. GHOSH & J. C. WHEELER. 1982. Astrophys. J., Lett. **260:** L83.
55. FAULKNER, J., D. N. C. LIN & J. PAPALOIZOU. 1983. Mon. Not. R. Astron. Soc. **205:** 359.
56. SMAK, J. 1984. Publ. Astron. Soc. Pac. **96:** 5.

ADDITIONAL READINGS

BASRI, G. & C. BERTOUT. 1989. Astrophys. J. **341:** 340.
GOLDREICH, P. & S. TREMAINE. 1982. Ann. Rev. Astron. Astrophys. **20:** 249.
KENYON, S. J., L. HARTMANN & R. HEWETT. 1988. Astrophys. J. **325:** 213.
LARSON, R. 1990. Mon. Not. R. Astron. Soc. In press.
LIN, D. N. C. 1980. Astrophys. J. **246:** 972.
LIN, D. N. C. & K. R. BELL. 1990. *In* Nonlinear Astrophysics. R. Buchler, Ed. In press.
LIN, D. N. C. & J. C. B. PAPALOIZOU. 1979. Mon. Not. R. Astron. Soc. **186:** 799.
LÜST, R. 1952. Z. Naturforsch. Teil A **7a:** 87.
PAPALOIZOU, J. C. B., J. FAULKNER & D. N. C. LIN. 1983. Mon. Not. R. Astron. Soc. **205:** 487.
PAPALOIZOU, J. C. B. & D. N. C. LIN. 1985. Astrophys. J. **285:** 818.
PRINGLE, J. E. 1981. Ann. Rev. Astron. Astrophys. **19:** 135.
SAWADA, K., T. MATSUDA & I. HACHISU. 1986. Mon. Not. R. Astron. Soc. **219:** 75.
SAWADA, K., T. MATSUDA, M. INOUE & I. HACHISU. 1987. Mon. Not. R. Astron. Soc. **224:** 307.
SMAK, J. 1969. Acta Astron. **19:** 155.
SPRUIT, H. C., T. MATSUDA, M. INOUE & K. SAWADA. 1987. Mon. Not. R. Astron. Soc. **229:** 517.
TOOMRE, A. 1981. *In* Structure and Evolution of Normal Galaxies, S. M. Fall and D. Lynden-Bell, Eds.: 111. Cambridge University Press. Cambridge, England.

Hydrodynamical Simulations of Collisions between Interstellar Clouds[a]

JOHN C. LATTANZIO

Institute of Geophysics and Planetary Physics
Lawrence Livermore National Laboratory
Livermore, California 94550

INTRODUCTION

The hydrodynamical evolution of self-gravitating gas clouds is a fundamental interaction in much of astrophysics. Collisions (I prefer the term *interactions*) between interstellar clouds have been identified as a possible site of star formation.[1] Two models for the formation and evolution of giant molecular clouds are the gravitational instability model and the random collisional build-up model;[2] the outcome of collisions is a crucial input for both of these scenarios. From a more general view, such interactions are clearly involved in understanding the dynamics of the gas content of spiral galaxies.[3] Similar interactions occur in protogalaxy evolution, but on a different scale.[4-6] Clearly, an understanding of the dynamics involved in such collisions will have wide applications, and is an important problem.

BASIC PRINCIPLES

The calculation of interstellar cloud collisions is an inherently complex procedure: the interaction is nonlinear, viscous, gravitational, radiative, and involves complex atomic physics. But some insights may be gained by considering simple scaling arguments. It might be expected that in the high-density collision interface it would be easier to get gravitational instabilities, but this is not necessarily the case. Consider, for example, a collision between two identical slabs of density ρ_0 and thickness d_0, so as to mimic a one-dimensional collision. If the relative velocity is $v = Mc$, where c is the local sound speed and M is the Mach number of the collision, then the collision produces a dense inner slab of density $\rho_1 = \rho_0 M^2$ with a thickness $d_1 \sim 2d_0/M^2$. Now let $\lambda_{J_0} = (\pi c^2/G\rho_0)^{1/2}$ be the Jeans length in the original cloud, and $\lambda_{J_1} = (\pi c^2/G\rho_1)^{1/2}$ be that in the compressed cloud. Parallel to the collision axis the length scale is now $d_1 = 2d_0/M^2$, so that in this direction there are $\lambda_{J_1}/d_1 = (M/2)(\lambda_{J_0}/d_0)$ Jeans lengths in the compressed cloud. But transverse to the collision axis the length scale is still d_0, so there we have

$$\frac{\lambda_{J_1}}{d_0} = \left(\frac{\lambda_{J_0}}{M}\right)\left(\frac{1}{d_0}\right) = \frac{1}{M}\left(\frac{\lambda_{J_0}}{d_0}\right).$$

[a]This work was performed under the auspices of the U.S. Department of Energy by the Lawrence Livermore National Laboratory under Contract W-7405-ENG-48.

158

We see that the supersonic collision actually stabilizes the clouds against gravitational collapse in the direction of the collision, whilst enhancing gravitational instability perpendicular to this. In fact, the ratio

$$\left(\frac{\lambda_J}{d}\right)_\perp \bigg/ \left(\frac{\lambda_J}{d}\right)_\parallel \sim M^{-2}.$$

Consider now two spherical clouds of mass M, radius R, and density ρ_0 that collide head-on to produce a disk. The surface density σ for the disk will vary with position in the disk, but we may take its average value as $8/3\rho_0 R$. Then the critical wavelength for the disk is[7]

$$\lambda_J^d = \frac{c^2}{G\sigma},$$

if the disk is isothermal. For an isothermal sphere

$$\lambda_J^s = \left(\frac{\pi c^2}{G\rho_0}\right)^{1/2} = qR,$$

and thus

$$\frac{\lambda_J^d}{\lambda_J^s} = \frac{3}{8\pi} q.$$

We infer from this that if the original cloud is far from unstable ($q \gg 1$), then the disk will be even further from instability. If, however, the cloud is unstable ($q < 1$), then the disk will be even more unstable. In practice the high-density regions produced by the collisions are never at rest and waiting for instabilities to develop. They are in motion, and should make us skeptical of trusting these simple arguments.

Real clouds, of course, are not isothermal. The cooling timescale in these clouds is very short, and any temporary high temperature seen in the shock will rapidly decrease through radiation from any molecules present. In fact, the cooling can be so efficient that the final temperature can be lower than the initial temperature (especially if the initial gas is atomic).[5,8]

THE NUMERICAL METHOD

Numerical models of cloud collisions were first constructed in one-dimension (infinite slabs) by Stone,[8] Smith,[9] and Struck-Marcell.[5] Models in two-dimensions were also constructed by Stone,[10] as well as Chieze and Lazareff,[11] and later by Gilden.[12] The first attempt to model collisions in three-dimensions was by Hausman.[13] The main failing of this calculation, using a simple finite-particle method, was the unphysical interpenetration of the particles (i.e., gas streams).

Lattanzio et al.[14] applied the smoothed particle hydrodynamics method (SPH[15,16]) to the problem of cloud collisions. Very briefly, in the SPH formalism an ensemble of "particles" (actually moving interpolation points) is used to approximate an arbitrary function $f(\vec{r})$ via

$$f(\vec{r}) = \sum_{j=1}^{N} \frac{f_i}{\rho_j} m_j W(|\vec{r} - \vec{r}_j|, h),$$

where $f_j = f(\vec{r}_j)$, m_j is the mass of particle j, ρ_j is the density at r_j, and the sum is over all N particles. The function W is the interpolation kernel, and is subject to the normalization condition $\int W d^3\vec{r} = 1$. The variable h is the smoothing length, and represents the typical resolution scale of the calculation. Usually, $W \to 0$ over $\sim 2h$, so the sums are not as computationally expensive as they first appear. Obviously, one prefers a spherically symmetric W. A simple Gaussian is not very good, and preference should be given to the spline-based kernels.[16]

One may apply this formalism to the equations of hydrodynamics. The continuity equation becomes simply $dm_i/dt = 0$. The density $\rho_i = \rho(r_i)$ is found from

$$\rho_i = \sum_{j=1}^{N} m_j W_{ij},$$

where $W_{ij} = W(|\vec{r}_i - \vec{r}_j|)$. For the momentum equation, after noting that

$$\vec{\nabla}\left(\frac{P}{\rho}\right) = -\frac{P}{\rho^2}\vec{\nabla}\rho + \frac{1}{\rho}\vec{\nabla}P,$$

we may write

$$\frac{d\vec{v}_i}{dt} = -\sum_{j=1}^{N} m_j \left(\frac{P_i}{\rho_i^2} + \frac{P_j}{\rho_j^2}\right) \vec{\nabla}_i W_{ij},$$

where $\vec{\nabla}_i W_{ij}$ is the gradient of $W(|\vec{r}_i - \vec{r}_j|)$ evaluated at r_i. This form conserves linear and angular momentum exactly (due to the symmetry of W_{ij} and the antisymmetry of $\vec{\nabla}_i W_{ij}$). To solve the streaming problem of Hausman, Lattanzio et al.[17] developed a new artificial viscosity V_{ij} that was added to the equation of motion. The equation used was

$$\frac{d\vec{v}_i}{dt} = -\sum_{j=1}^{N} m_j \left(\frac{P_i}{\rho_i^2} + \frac{P_j}{\rho_j^2}\right) (1 + V_{ij})\vec{\nabla}_i W_{ij},$$

where $V_{ij} = -\alpha\mu_{ij} + \beta\mu_{ij}^2$ and

$$\mu_{ij} = \begin{cases} 0, & \text{if } \vec{v}_{ij} \cdot \vec{r}_{ij} > 0, \\ \dfrac{h}{c}\dfrac{\vec{v}_{ij} \cdot \vec{r}_{ij}}{r_{ij}^2 + 0.1h^2}, & \text{otherwise;} \end{cases}$$

α and β are constants of order unity, and we have used the notation $\vec{x}_{ij} = \vec{x}_i - \vec{x}_j$.

With this form of the equation of motion the viscous forces conserve linear and angular momentum exactly, are Galilean invariant, and vanish for rigid rotation. (Note that this form of viscosity has both shear and bulk components.) It has been extensively tested,[17] and prevents particle interpenetration for $M \leq 50$, with the extent of the spread of the shock determined by α and β. In all calculations to be discussed here we have used $\alpha = 2$ and $\beta = 4$, which spreads the discontinuty over about $3h$, and is thus quite acceptable.

To solve Poisson's equation it is often best to eschew particle methods because of their prohibitive cost (for direct summation; the recently developed "tree" structures are an exception[18]). Rather, we use a grid-based method[16,19] of resolution h. The particular Poisson solver used in the calculations to be described below has been continually updated from Lattanzio et $al.$[14] (hereafter Paper 1), Lattanzio and Henriksen[20] (hereafter Paper 2), and Keto and Lattanzio[21] (hereafter Paper 3; see Monaghan and Lattanzio[22]). Briefly, the method is the following. A cubic grid is placed over the particle distribution, and the density is determined at the grid vertices. The potential on the surface of the cube is obtained by multipole expansions. We solve for the potential on the grid by Gauss–Seidel iterations. To speed the convergence we use multiple grids of decreasing resolution, which enable rapid solution of the long wavelength variations in the gravitational potential. In Paper 1 we used two grids, with a subgrid correction to obtain an accuracy of order h^2. Paper 2 used three grids and fourth-order finite-difference formulas. In Paper 3 we used four grids and included a nested four grids that are triaxial rather than cubic, providing higher resolution for a given memory constraint.[22] Papers 1–3 should be consulted for details.

RESULTS

From now on we restrict the discussion to the isothermal equation of state. We start by discussing the calculations of Paper 1, which studied collisions with equal and unequal masses, and impact parameter b both zero and nonzero. FIGURE 1 shows a collision between two clouds with $\alpha = E_{th}/|E_{grav}| = 5.3$, so the clouds are initially far from gravitational instability. The Mach number of the collision is 16.6. We see here all the essential features expected from the simplified arguments given earlier: rapid compression, spreading of the gas perpendicular to the collision direction, a pressure-driven reexpansion, and a tendency toward focusing of material perpendicular to the collision direction during the subsequent reexpansion. All of these features were also seen in the two-dimensional calculations.[10–12] FIGURE 2 shows a qualitatively similar case, with $\alpha = 1.3$ and $v/c = 8.3$. Here we indeed see gravitational instability, triggered by the contraction perpendicular to the collision direction, which occurs during the reexpansion. Note that much matter is lost from the system, again in agreement with the earlier two-dimensional calculations. FIGURE 3 shows a head-on impact between two clouds of the same initial density, but mass ratio 2.5, the larger cloud having $\alpha = 1.3$ and $v/c = 8.3$. Despite the focusing of matter in the transverse direction, the collision destroys the small cloud and disrupts the larger. In general, the results of Paper 1 for head-on collisions indicate that if the two clouds are initially near instability, then gravitational collapse can result, provided that both the mass ratio of the two clouds and the relative velocity are not "too large": $M_2/M_1 \leq 3$ and $v/c \leq 6$. These results are in agreement with the higher resolution two-dimensional calculations of Gilden.[12]

Paper 1 also considered off-center collisions of clouds. FIGURE 4 shows two clouds with $\alpha = 1.3$ colliding with $v/c = 16.6$ and impact parameter $b = R$, the initial

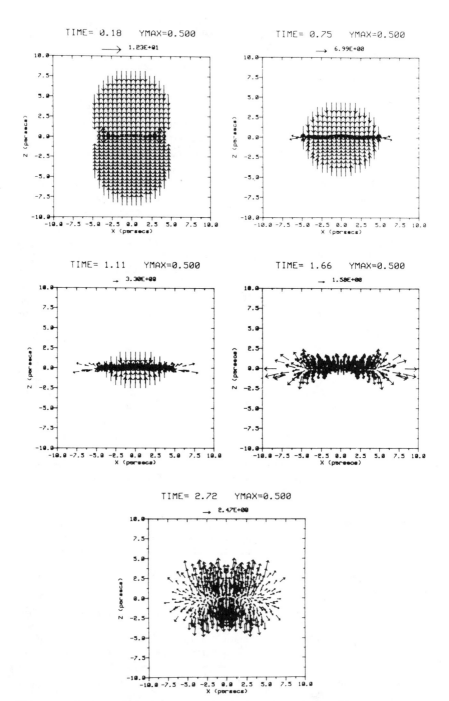

FIGURE 1. Head-on collision between two clouds of equal mass $361M_\odot$ with $R = 5$ pc and $v/c =$ 16.6. The times quoted are in 10^6 years ($t_{ff} = 9.7 \times 10^6$ years). Velocities are in km s^{-1}, and we plot only those particles with $|y_i| <$ YMAX. All lengths are in parsecs. (Reprinted with permission from the Royal Astronomical Society.)

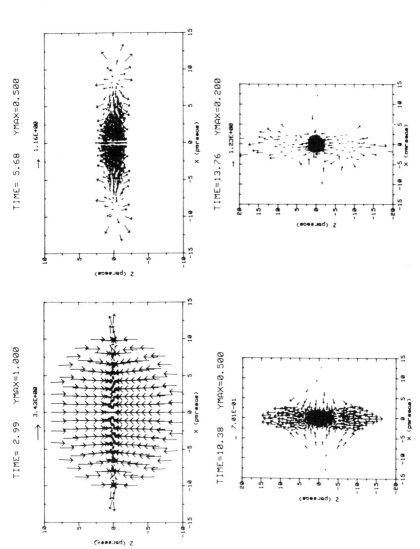

FIGURE 2. Same as FIGURE 1, but for $M = 2887M_\odot$, $R = 10$ pc, and $v/c = 8.3$. (Reprinted with permission from the Royal Astronomical Society.)

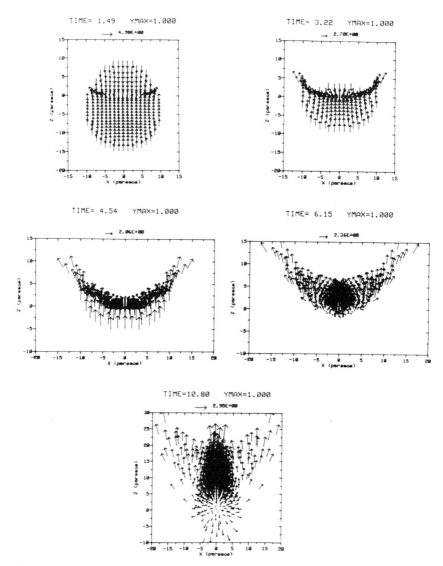

FIGURE 3. Same as FIGURE 2, but with $M_2/M_1 = 2.5$ and $M_2 = 2887M_\odot$. (Reprinted with permission from the Royal Astronomical Society.)

cloud radius. We see that the unhindered material in the clouds is essentially unaffected by the collision, whilst the overlapping region disperses during the reexpansion phase following maximum compression. The nonzero value of b has added some (orbital) angular momentum to the system, which can be seen in the motion of the departing components. By reducing v/c to 1.04 we find that the resulting configuration is far less dissipative, and FIGURE 5 shows that a relatively stable, rotating bar forms. These two examples, plus the many other cases in Paper 1, highlight the difficulties associated with estimating the behavior of ensembles of interacting clouds.[4] A wide variety of outcomes are seen, depending on v/c, b, and M_2/M_1.

The addition of rotational support in the form of spin angular momentum was

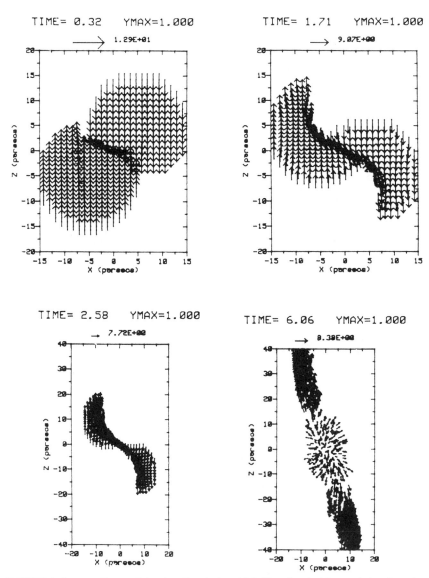

FIGURE 4. Same as FIGURE 2, but $b = R$ and $v/c = 16.6$. (Reprinted with permission from the Royal Astronomical Society.)

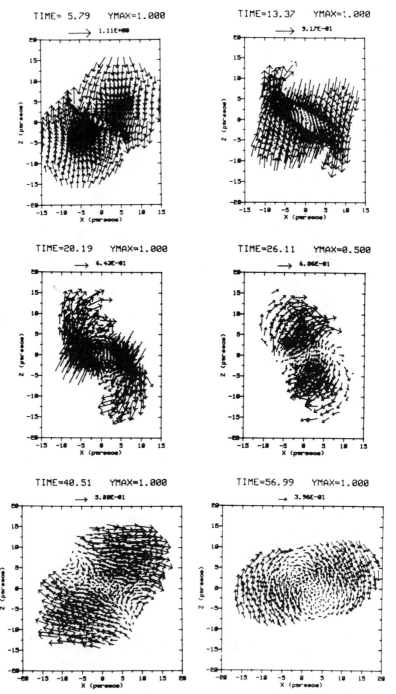

FIGURE 5. Same as FIGURE 4, but $v/c = 1.04$. (Reprinted with permission from the Royal Astronomical Society.)

considered in Paper 2. This greatly extends the parameter space to be examined. Paper 2 considered only the restricted space where the spin and orbital angular momenta are parallel (or antiparallel), and the clouds have equal masses (with $\alpha = 0.55$), and solid body rotation. FIGURE 6 shows one example from Paper 2. Here the collision is head-on, $\beta = E_{rot}/|E_{grav}| = 0.073$ for each cloud, and $v/c = 1$, with the spins parallel. Despite dissipation at the interface, the large angular momentum of this configuration leads to a collapsing lump with matter in near-Keplerian orbits on the outside. FIGURE 7 shows a similar case, except $\beta = 0.292$ and the spins are

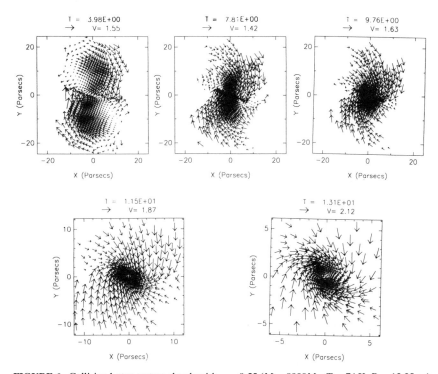

FIGURE 6. Collision between two clouds with $\alpha = 0.55$ ($M = 8000M_\odot$, $T = 74$ K, $R = 12.25$ pc) and $\beta = 0.073$ (with solid body rotation). Here $t_{ff} = 7.9 \times 10^6$ years, $v/c = 1$, and the spins are parallel. The unit of time is 10^6 years, and the velocities are in km s^{-1}. (Reprinted with permission from the Royal Astronomical Society.)

initially antiparallel, so that there is *no* net angular momentum. Note that we clearly see a "unipolar outflow" in the resulting configuration. In general, for the head-on collisions, we see that the total angular momentum is the prime controller of the outcome. For the cases with no net angular momentum almost all result in gravitational collapse, regardless of v/c and β. For the clouds with the spin angular momenta parallel, however, only for $\beta \lesssim 0.3$ and $v/c \lesssim 5$ do we see gravitational instability. We do find that some cases form rapidly rotating bars that could act as gravitational torques for the redistribution of angular momentum.[23] We also find that the mutual gravitational acceleration of the clouds often overcomes the shock dissipation, in the sense that the total kinetic energy increases during the interaction.

By taking $b \neq 0$ we are adding some orbital angular momentum to the system. FIGURE 8 shows the case where the cloud spins are mutually parallel and parallel to

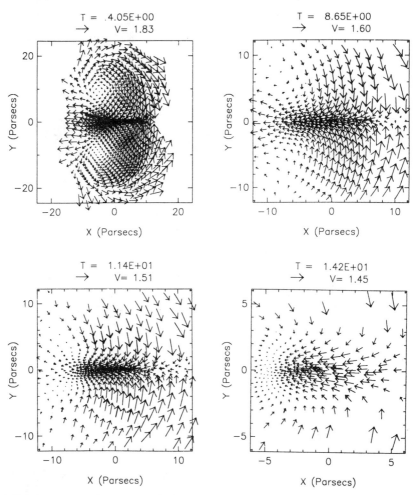

FIGURE 7. Same as FIGURE 6, but with $\beta = 0.292$ and antiparallel spins. (Reprinted with permission from the Royal Astronomical Society.)

the orbital angular momentum, with each cloud having $\beta = 0.073$. The relative velocity is $v = 4c$ and $b = R$. We see that the transfer between spin and orbital angular momentum is crucial to the outcome of the collision: If enough angular momentum can be bound into orbital motion, then the core is free to collapse. This is

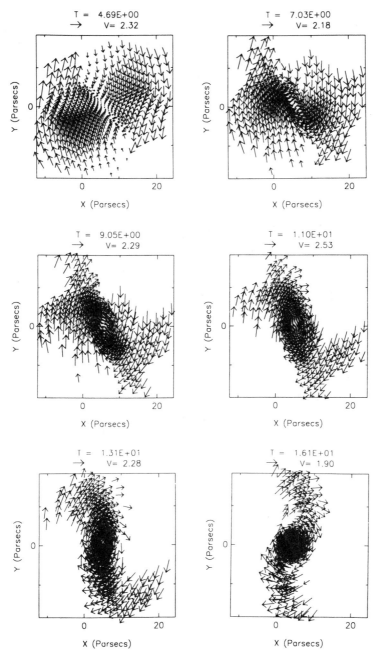

FIGURE 8. Same as FIGURE 6, but with $v/c = 4$ and $b = R$. (Reprinted with permission from the Royal Astronomical Society.)

what is seen in the figure. FIGURE 9 shows the same case, but with v/c doubled to 8. Here we see the initial compression zone torn into two clouds, much like as shown in FIGURE 4. Finally, FIGURE 10 shows a case where the two spin angular momenta are antiparallel; otherwise, it is the same as shown in FIGURE 8. In this case, we see the formation of one tail only, which contains all of the final orbital angular momentum of the system.

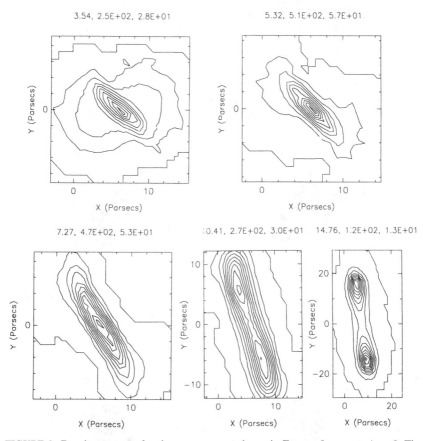

FIGURE 9. Density contours for the same case as shown in FIGURE 8, except $v/c = 8$. The numbers above each panel are the times (in 10^6 years), the value of the maximum density contour and the level between successive contours (in cm^{-3}). (Reprinted with permission from the Royal Astronomical Society.)

For the restricted symmetries considered in Paper 2 it was shown that a reasonably good predictor of collapse could be derived from the initial moment of inertia of the interacting configuration. For (initially) spherical clouds of mass M and radius R the moment of inertia about the center of mass is

$$I_{cm} = 2 \left(\tfrac{2}{5} MR^2 + M(b/2)^2 \right)$$

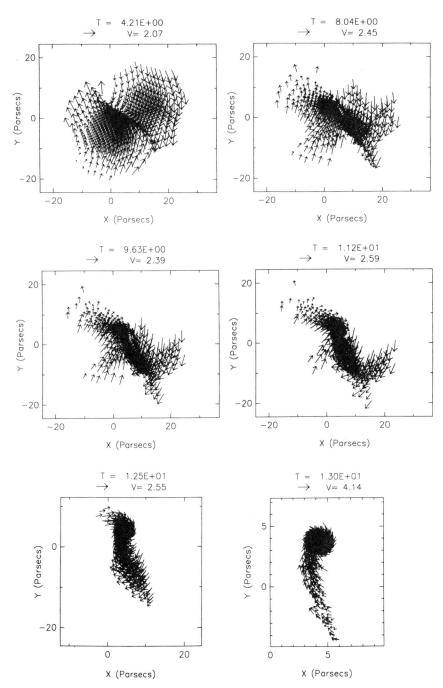

FIGURE 10. Same case as shown in FIGURE 8, except for the antiparallel spins. (Reprinted with permission from the Royal Astronomical Society.)

for impact parameter b. The angular momentum is

$$L_{cm} = Mv\frac{b}{2} + \frac{4}{5}M\Omega R^2 k,$$

where

$$k = \begin{cases} 0, & \text{for antiparallel spins,} \\ +1, & \text{for parallel spins, parallel to the orbital angular momentum,} \\ -1, & \text{for parallel spins, antiparallel to the orbital angular momentum.} \end{cases}$$

Here each cloud has a rotational frequency Ω and the relative velocity is v. Defining Ω_0 as the rotation that generates an equatorial velocity of c (i.e., $c = R\Omega_0$), we have

$$\Omega_0 = \frac{1}{t_{ff}}\sqrt{\frac{\alpha\pi^2}{20}},$$

where $t_{ff} = (3\pi/32G\rho)^{1/2}$ is the free-fall timescale of the individual clouds, and α is the ratio of the thermal to gravitational energy of each cloud (with temperature T):

$$\alpha = \frac{5}{2}\frac{\Re TR}{\mu GM},$$

where $\Re$ is the gas constant and μ is the mean molecular weight. Then a natural angular frequency in the center-of-mass frame Ω_{cm} is given by

$$Q = \frac{\Omega_{cm}}{\Omega_0} = \frac{|L_{cm}|}{I_{cm}\Omega_0} = \frac{|(\Omega/\Omega_0)k + \frac{5}{8}(v/c)(b/R)|}{1 + \frac{5}{8}(b/R)^2}.$$

Paper 2 found that $Q = \Omega_{cm}/\Omega_0 \simeq 2.5$ divides the dispersing interactions ($Q \geq 2.5$) from those resulting in gravitational instability ($Q \leq 2.5$). (Because Ω_0 is a function of t_{ff} and α, we can express the critical Ω_{cm} in terms of these parameters, if desired.)

The lifetime of the interacting pair was also discussed in Paper 2, and found to be quite sensitive to the angle that the relative velocity makes with the head-on direction at the first point of contact (see Paper 2 for details). In any event, long-lived structures are quite common, due to the rotational support, in agreement with the basic ideas of Pumphrey and Scalo.[4]

COMPARISON WITH OBSERVATIONS

Note that Paper 2 referred to outflows or "collimated splashes," and predicted the appearance of unipolar and bipolar flows with velocities appropriate to the velocity dispersion of the individual clouds. Such flows would be present in the absence of a driving (embedded) energy source, such as a young star. This has indeed been observed by Blitz et al.[24] (hereafter BMW). Keto and Lattanzio[21] (Paper 3) undertook a quantitative investigation of this model.

They took parameters appropriate to high-latitude clouds and ran two collisions—one with $b = 0$ and one with $b = R$. In each case the masses are equal at $100M_\odot$, with $R = 1.1$ pc, $T = 80$ K, $\Omega/\Omega_0 = 0.85$, and $v = 4$ km s$^{-1} = 5.8c$. FIGURE 11 shows the

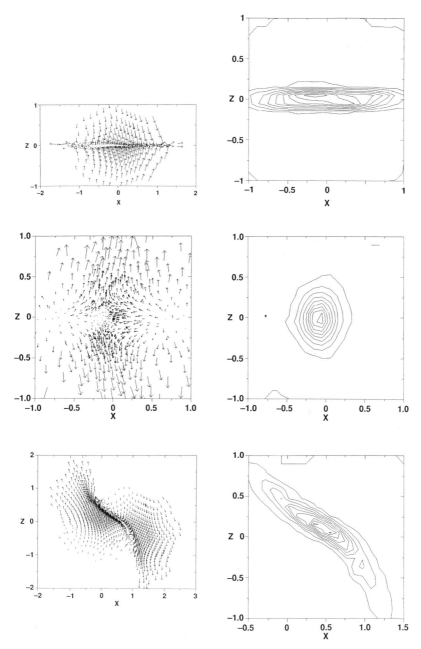

FIGURE 11. Three snapshots from the two collisions considered in Paper 3. The upper four are for the $b = 0$ case (see text) at $t/t_{ff} = 0.34$ (*top*) and $t/t_{ff} = 0.87$ (*center*). The lower two are from the $b = R$ case at $t/t_{ff} = 0.40$. The length unit is 1.1 pc, and the contour interval is 2.0×10^{-21} g cm^{-3} (*top*), 4.9×10^{-21} g cm^{-3} (*center*), 1.7×10^{-21} g cm^{-3} (*bottom*). The largest velocity vectors correspond (approximately) to 1.6 km s^{-1} (*top*), 0.82 km s^{-1} (*center*), 2.9 km s^{-1} (*bottom*).

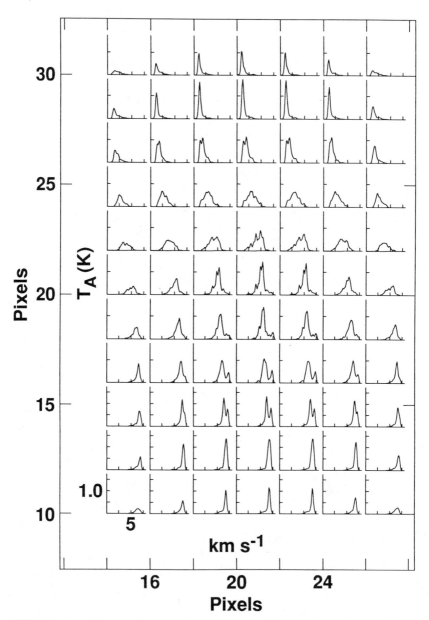

FIGURE 12. A set of spectra from the orientation and collision shown at the bottom of FIGURE 11.

three "snapshots" taken from these two calculations. For each of these snapshots six representative orientations on the sky are chosen (see Paper 3 for details). The calculated density and velocity field is then used in a three-dimensional radiative transfer calculation[25] to determine synthetic ^{13}CO spectra at various positions through the configurations. FIGURE 12 shows a typical case. Note the rich variety of spectra

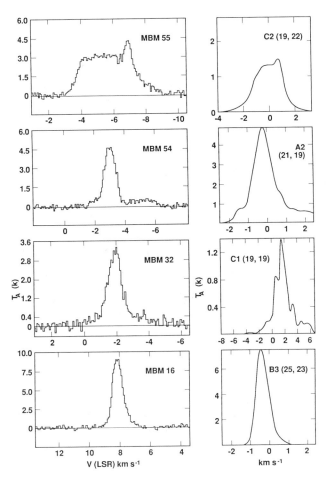

FIGURE 13. ^{12}CO spectra from BMW (*left*) and the ^{13}CO synthetic spectra from the models (*right*).

seen, including single lines, obvious outflows, double lines, and other complex phenomena. After examining the calculations we have been able to find calculated spectra that model those of BMW exceedingly well, as shown in FIGURE 13. This is powerful evidence for interstellar collisions, showing them to be frequent and observable events in the interstellar medium (see further evidence quoted in Paper

3). It was also found (see Paper 3) that star formation could be initiated in these high-latitude clouds by collisions, provided the impact parameter was small (see also Paper 1).

Using this powerful combination of three-dimensional hydrodynamics and radiative transfer, it is possible to use observations of cloud complexes as diagnostics for determining their dynamical behavior. It is also strong evidence for the validity of the calculations.

CONCLUSIONS

We have summarized the most recent three-dimensional calculations of interacting ("colliding") interstellar clouds. The collisions exhibit a rich variety of "final" outcomes. Many of the resulting structures, if viewed without knowledge of the prior history, are not obviously identified with "collisions" between entities called "clouds." This should be remembered when interpreting observations. We imagine the interstellar medium as a complex of interacting fluids with various density and velocity fields, rather than consisting of individual building blocks called clouds.

The calculations show that it is possible for collisions to induce star formation, but that a very common result is the disruption of the interacting clouds. For a collision to lead to gravitational instability seems to require roughly head-on collisions between nearly equal mass clouds at low relative velocity, and that the clouds themselves not be too far from gravitational instability initially.

In any event, it is now possible, through a combination of three-dimensional hydrodynamics and radiative transfer, to use observed spectra as probes of the fully three-dimensional structure and motion of the interstellar medium. Indeed, it is even possible to directly simulate complexes of gas clouds,[26] and this will certainly give us new understanding into the structure and evolution of the interstellar medium, including the processes involved in turning this gaseous medium into new stars.

ACKNOWLEDGMENTS

I would like to thank my collaborators on Papers 1–3: Dick Henriksen, Eric Keto, Helen Pongracic, and Phil Schwarz, for their contributions. A special thanks to Joe Monaghan for his continued interest and enthusiasm, as well as comments on the manuscript. I also thank John Scalo for useful discussions.

REFERENCES

1. SCOVILLE, N. Z., D. B. SANDERS & D. P. CLEMENS. 1986. Astrophys. J., Lett. **310:** L77.
2. ELMEGREEN, B. G. 1989. Preprint.
3. ELMEGREEN, B. G. & D. M. ELMEGREEN. 1989. *In* Evolutionary Phenomena in Galaxies, J. Beckman and B. Pagel, Eds. Cambridge University Press. Cambridge, England. In press.
4. PUMPHREY, W. & J. M. SCALO. 1983. Astrophys. J. **269:** 531.
5. STRUCK-MARCELL, C. 1982. Astrophys. J. **259:** 116.
6. ———. 1982. Astrophys. J. **259:** 127.

7. LARSON, R. B. 1985. Mon. Not. R. Astron. Soc. **214:** 379.
8. STONE, M. E. 1970. Astrophys. J. **159:** 277.
9. SMITH, J. A. 1980. Astrophys. J. **238:** 842.
10. STONE, M. E. 1970. Astrophys. J. **159:** 293.
11. CHIEZE, J. P. & B. LAZAREFF. Astron. Astrophys. **91:** 290.
12. GILDEN, D. L. 1984. Astrophys. J. **279:** 335.
13. HAUSMAN, M. A. 1981. Astrophys. J. **245:** 72.
14. LATTANZIO, J. C., J. J. MONAGHAN, H. PONGRACIC & M. P. SCHWARZ. 1985. Mon. Not. R. Astron. Soc. **215:** 125.
15. GINGOLD, R. A. & J. J. MONAGHAN. 1977. Mon. Not. R. Astron. Soc. **181:** 375.
16. MONAGHAN, J. J. & J. C. LATTANZIO. 1985. Astron. Astrophys. **149:** 135.
17. LATTANZIO, J. C., J. J. MONAGHAN, H. PONGRACIC & M. P. SCHWARZ. 1985. SIAM J. Sci. Stat. Comput. **7:** 591.
18. HERNQUIST, L. & N. KATZ. Astrophys. J., Suppl. Ser. **70:** 419.
19. MONAGHAN, J. J. 1985. Comput. Phys. Rep. **3**(2): 71.
20. LATTANZIO, J. C. & R. N. HENRIKSEN. 1988. Mon. Not. R. Astron. Soc. **232:** 565.
21. KETO, E. R. & J. C. LATTANZIO. 1989. Astrophys. J. **346:** 184.
22. MONAGHAN, J. J. & J. C. LATTANZIO. 1990. Astrophys. J. In press.
23. HENIKSEN, R. N. & B. E. TURNER. 1984. Astrophys. J. **287:** 200.
24. BLITZ, L., L. MAGNANI & A. WANDEL. 1989. Astrophys. J., Lett. **331:** L127.
25. KETO, E. R. 1989. Submitted for publication in Astrophys. J.
26. MONAGHAN, J. J. & S. R. VARNAS. 1988. Mon. Not. R. Astron. Soc. **231:** 515.

Galaxy Formation: Gas Dynamics Versus Stellar Dynamics

G. CONTOPOULOS,[a,b] N. VOGLIS,[b] AND N. HIOTELIS[b]

[a]Department of Astronomy
University of Florida
Gainesville, Florida 32611

[b]Department of Astronomy
University of Athens
Athens, Greece

INTRODUCTION

The formation of galaxies takes place in an environment consisting of all the other galaxies of the Universe. Its evolution depends not only on its internal dynamics but also on the influence of the other galaxies. If the galactic environment has a certain degree of asymmetry, it applies a torque on a particular galaxy, which is proportional to

$$\int \frac{4\pi R^2 \rho \, dR}{R^3} = \int \rho \frac{dR}{R} ,$$

where ρ is the density of galaxies at distance R. Therefore the effect of distant galaxies is important and cannot be ignored in general. It is only the large-scale isotropy of the Universe that avoids the logarithmic divergence of the total torque.

Another argument leading to the same result is the observed fact that nearby galaxies do not have well-anticorrelated spins,[1,2] as one would expect if the rotation of each galaxy is due mainly to its closest neighbors.

The effects of the environment of a galaxy have been studied by Quinn, Salmon, and Zurek,[3] Barnes and Efstathiou,[4] and by Zurek, Quinn, and Salmon.[5] A recent systematic study of these effects, starting essentially at recombination time, was made by Voglis and Hiotelis[6] using N-body simulations. The galaxy is represented by 1072 points, forming initially an ellipsoidal configuration, and it is surrounded by 272 other galaxies forming a much larger system, which has an asymmetric perturbation with a different orientation, representing one or more clusters of galaxies. Both the galaxy and the clusters initially follow the expansion of the Universe, but with a slower rate. The galaxy reaches a maximum expansion, and then collapses. The same happens to the clusters at a later time.

During the expansion phase and until the collapse of the minor axis of the galaxy the effects of dissipation, due to gas, are insignificant. Thus, the evolution of the galaxy is well represented by an N-body simulation.

After the collapse of the minor axis the N-body evolution differs significantly from that of the gas.

In the following sections we describe separately the stellar-dynamical and gas-dynamical evolution of a galaxy, and then we compare the two. Finally, we discuss the results of the gas-dynamical calculations.

STELLAR-DYNAMICAL EVOLUTION

The overall stellar evolution of a galaxy inside a cluster is represented in FIGURE 1. We give the time evolution of the angular momentum of eight successive shells of the galaxy containing 3/10, 4/10, . . . , 10/10 of the total mass. We see that during the expansion phase the outer shells acquire most of the angular momentum, and during the collapse phase angular momentum is transferred to the inner shells. Later on, however, the continuing effect of the tidal field of the other galaxies, as well as the interaction of the various parts of the galaxy among themselves, produce a redistribution of the angular momentum. The material of the outermost shells expands to large distances and escapes from the galaxy. The angular momentum of the inner shells is reduced from their maximum value, and after a transition period it tends to a roughly stationary distribution that does not change significantly any more. In other cases, in which the timescale of the collapse of the clusters is of the same order as the collapse time of the galaxy itself, the angular momentum of the galaxy may reverse its sign, or the outer shells may rotate in a direction opposite to that of the inner shells.

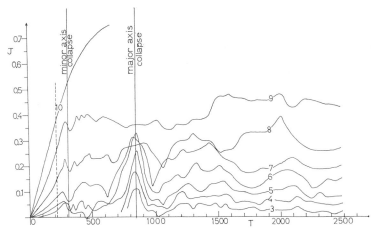

FIGURE 1. Growth and distribution of the angular momentum of a stellar galaxy inside an environment of clusters. We give the time evolution of the angular momentum of 8 successive shells of the galaxy containing 3/10, 4/10, . . . , 10/10 of the total mass. The times, when the collapses of the minor and the major axes occur, are shown. The angular momentum grows during the expansion phase, and it is transferred to the inner shells during the collapse of the major axis. Later on, the angular momentum reaches a semipermanent distribution. The *dashed line* at time $T = 220$ shows the moment when we start the parallel evolution of the gaseous galaxy.

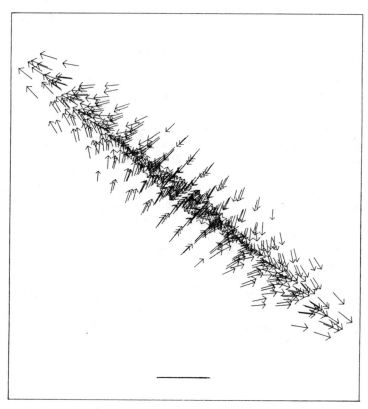

FIGURE 2. The velocity field of the protogalaxy at $T = 220$. This configuration is chosen as initial conditions for the evolution of the gaseous galaxy. (Time = 220; YZ-plane; bar length = 20.)

A more detailed picture of the evolution of a stellar galaxy is given by the velocity field of the galaxy at successive times (FIGS. 2–5). The unit of length in these figures depends on the epoch of galaxy formation. It is ≈ 2 kpc if galaxies form at redshift $z \approx 5$. The unit of time is then about 1 My.

As initial conditions for the comparison of stellar and gas dynamics we take the positions and velocities a little before the collapse of the minor axis (dashed line of FIG. 1 at $T = 220$). At that time the major axis is still expanding, while a maximum of the density appears along the plane of symmetry of the galaxy (FIG. 2). A little later, however, the expansion of the major axis stops, while most stars cross the plane of symmetry and move outward (FIG. 3; $T = 373$).

After that stage the main feature of the stellar component is a large degree of irregular motions (heating). This is clear already at $T = 681$ (FIG. 4), but it is most prominent at later times (e.g., at $T = 2305$, FIG. 5). The N-body galaxy takes an almost stationary elliptical configuration that rotates slowly inside an extended

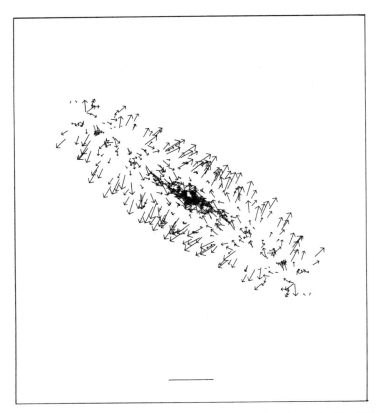

FIGURE 3. The stellar galaxy a little after the collapse of the minor axis ($T = 373$). The expansion of the major axis has stopped. Many stars have passed through the plane of symmetry moving outwards. (Time = 373; YZ-plane; bar length = 20.)

three-axial halo. It is pressure supported, due to the random motions of the stars. Such a model is applicable to elliptical galaxies, but cannot represent disk galaxies. In fact, the observed disk galaxies have a much larger rotational energy in comparison with the energy of random motions than in purely stellar-dynamical models.

GAS-DYNAMICAL EVOLUTION

Very different is the evolution of a galaxy that is mainly gaseous, that is, when dissipation plays a major role in its development. The gas simulation was done with a soft-particle hydrodynamical code (SPH), analogous to that used by Monaghan and Lattanzio.[7]

In this code the fluid is composed of elements that are represented by "smooth"

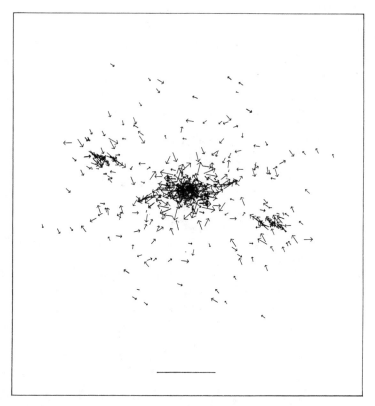

FIGURE 4. The stellar galaxy at $T = 681$. The central part along the major axis has collapsed, and there is some secondary infall. The main feature, however, is a large degree of irregular motions. (Time = 681; YZ-plane; bar length = 20.)

particles. Every particle is smoothed up to a radius $2h$ by a kernel

$$W(r, h) = \frac{1}{\pi h^3} \begin{cases} 1 - \tfrac{3}{2}x^2 + \tfrac{3}{4}x^3, & 0 \le x \le 1 \\ \tfrac{1}{4}(2 - x)^3, & 1 \le x \le 2 \\ 0, & x \ge 2 \end{cases}$$

with $x = r/h$. The density ρ_i at the location of the particle i is then given by

$$\rho_i = \sum_j m_j W(r, h),$$

where now $r = |r_i - r_j|$.

The equations of motion are

$$\frac{dv_i}{dt} = -\sum_j m_j \left(\frac{P_i}{\rho_i^2} + \frac{P_j}{\rho_j^2} \right) \left(1 + \prod_{ij} \right) \nabla_i W_4 - \nabla_i \Phi_i.$$

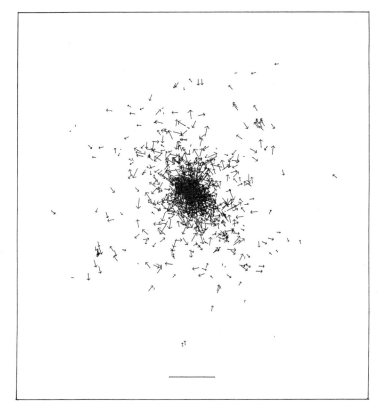

FIGURE 5. The stellar galaxy at $T = 2305$. The N-body galaxy takes the form of a slowly rotating elliptical configuration inside an extended three-axial halo. It is supported by the pressure of random motions. (Time = 2305; YZ-plane; bar length = 20.)

In these formulas m_j is the mass assigned to the particle j, P_j is the pressure of the fluid at the location of this particle, Φ_i is the potential at the position of the particle i, and Π_{ij} represents an artificial viscosity given by

$$\prod_{ij} = -\alpha\mu_{ij} + \beta\mu_{ij}^2,$$

with

$$\mu_{ij} = \frac{h\mathbf{v}_{ij} \cdot \mathbf{r}_{ij}}{c_{ij}(r_{ij}^2 + \eta^2)}$$

(c_{ij} is the average sound speed between the positions of the particles i and j, and the parameter $\eta^2 = 0.1h^2$ is a "softening parameter" introduced only to avoid numerical divergences; $\mathbf{r}_{ij}$ and $\mathbf{v}_{ij}$ are defined as $\mathbf{r}_{ij} = \mathbf{r}_i - \mathbf{r}_j$ and $\mathbf{v}_{ij} = \mathbf{v}_i - \mathbf{v}_j$).

The thermal energy per unit mass u_i at the position of the particle i is given by the equation

$$\frac{du_i}{dt} = \frac{1}{2} \sum_j m_j \left(\frac{P_i}{\rho_i^2} + \frac{P_j}{\rho_j^2}\right)\left(1 + \prod_{ij}\right) \mathbf{v}_{ij} \nabla_i W.$$

The system of equations closes with the equation of state

$$P_i = (\gamma - 1)\rho_i u_i,$$

where γ is the ratio of the specific heats. The gravitational force, $-\nabla_i \Phi_i$, is evaluated either by a grid method or by a particle–particle calculation.

The initial conditions are the same as those of the N-body stellar system, namely the configuration of FIGURE 2 that was derived by the N-body simulation a little before the total collapse of the minor axis. Up to that time the dissipation in the galaxy is insignificant. After the collapse of the minor axis, however, the effects of the dissipation are very important. In FIGURES 6–8 we give the form of the gaseous galaxy at the same times as for the stellar galaxy (FIGS. 3–5).

FIGURE 6 gives the form of the gaseous galaxy at $T = 373$ and shows an impressive difference from the corresponding stellar galaxy (FIG. 3). All the gas is concentrated

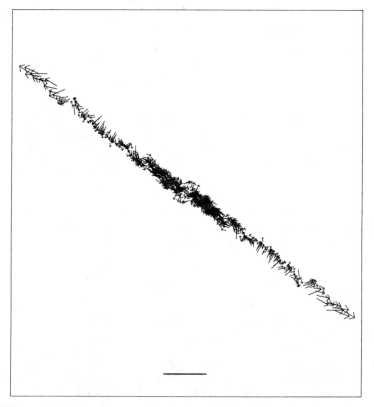

FIGURE 6. The gaseous galaxy at $T = 373$. All the gas is concentrated temporarily near a plane. (Time = 373; YZ-plane; bar length = 20.)

FIGURE 7. The gaseous galaxy at $T = 681$. The gas already shows a tendency toward regular motions. A strong shock is formed near the center on a plane roughly perpendicular to the direction of the collapsing major axis. (Time $= 681$; YZ-plane; bar length $= 20$.)

near a temporary "galactic plane." However, there are still significant motions that lead to further evolution of the galaxy. At time $T = 681$ (FIG. 7) the gas already shows a tendency toward regular motions. This is more strongly emphasized later, leading to almost circular motions (see, e.g., FIG. 8 at $T = 2305$). In this case the motions are highly organized. This should be contrasted with the stellar galaxy of FIG. 5, at the same time, which practically contains only irregular motions.

Thus, gas dynamics is essential in understanding the formation of disk galaxies, for example, for spiral galaxies.

DISCUSSION

Although the main result of the present study, namely the organization of the motions in a collapsing gaseous galaxy, is undoubtedly correct, the details of the calculations are not yet certain.

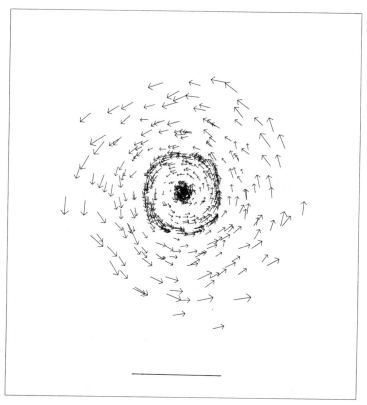

FIGURE 8. The gaseous galaxy at $T = 2305$ shows a highly organized motion. The fluid elements are separated into three groups. (**1**) The centers of $\approx 50\%$ of the fluid elements form a disk-like distribution near the center within a radius $h/2$ (*central dark region*). (**2**) A shell-like distribution with radius $2h$ contains $\approx 25\%$ of fluid element centers. The outer boundary of the shell is projected as a ring in the figure. (**3**) The rest $\approx 25\%$ of the material follows circular orbits. (Time = 2305; YZ-plane; bar length = 20.)

The main question that remains open is how large must the number of soft particles be and how small their size h in order to trust the results of the SPH code.

Durisen *et al.*[8] made a comparison of an SPH code with a finite difference hydrodynamical code. Both codes give similar results in general, but they do not agree in the details. Both have some advantages and disadvantages. One disadvantage of the smooth-particle code is its poor representation of the low-density regions.[8] This defect seems to have been remedied by a more recent code, developed by Hernquist and Katz,[9] using individual values of h for each particle. In order to reduce the calculation time, this code considers the potential of the distant particles by using a multipole expansion formula. Thus, the total time is only of $O(N)$ instead of the much longer times needed by other codes.

A more basic difficulty still persists, however, in all SPH codes. This refers to the crucial role played by the value of the smoothing radius h. This radius introduces a scale on which depends the final equilibrium configuration. We can separate the fluid

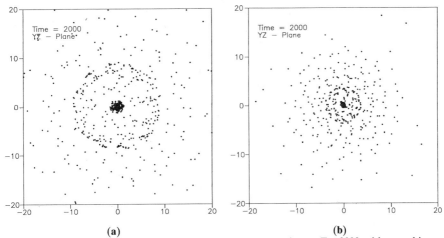

(a)   (b)

FIGURE 9. The effect of the smoothing length h. A gaseous galaxy at $T = 2000$, with smoothing length equal to (a) $h_1 = 5$ and (b) $h_2 = 2/5h_1$. The disk and shell-like groups are reduced in size by the same factor in (b).

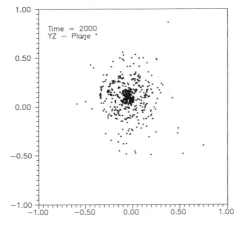

FIGURE 10. The central disk-like region of FIGURE 9(b) in detail. It contains a conspicuous spiral feature.

elements of the final configuration into three groups. The first group forms a disk-like distribution around the center of the galaxy, of radius $h/2$, containing more than 50 percent of the particles in our models (central dark region in FIG. 8). This group rotates almost as a solid body and is mainly pressure supported.

Around this group a second spheroidal shell-like distribution has a radius $2h$, and contains ~ 25 percent of the particles. It is also mainly pressure supported, and has an almost solid-body rotation, but with a lower angular velocity than the central part. A manifestation of this group is the ring formed in FIGURE 8, which is a projection of the outer boundary of the shell.

The rest of the mass is outside this ring, forming a large spheroidal configuration. This last group has large rotational velocities and is rotation supported. It is important, because it carries the main part of the total angular momentum. Its mass is small, however, of order 20–25 percent of the total mass.

In order to check the validity of our conclusion that the system separates into three groups, we repeated our calculations with a value of h reduced by a factor of 2/5. Our final results are shown in FIGURES 9(b) and 10 (at $T = 2000$). We see that the sizes of the central region and of the ring around it are reduced by the same factor of 2/5, in comparison with the previous case for the same time $T = 2000$ [FIG. 9(a)].

If we focus on the central region of FIGURE 9(b), we see an interesting structure, namely some conspicuous spiral forms (FIG. 10). All the particles, however, have a size of radius about h; therefore, the actual distribution of the mass is more extended than the distribution of their centers, given in FIGURE 10. Therefore, the details seen in this figure are not certain.

It seems that the spiral structure and all the details of the central part of the galaxy are a consequence of the shock produced during the collapse of the major axis of the galaxy. As regards the outer shell of FIGURES 8 and 9, it is due to the pressure developed in the central region that stops the infall of more matter toward the center.

Similar solid-body rotations have been found by other authors, for example, Lattanzio and Heriksen[10] in their simulation of off-center colliding clouds. This kind of evolution of a gaseous galaxy is to be expected, because of the viscosity of the gas, which tends to eliminate all relative motions of neighboring fluid elements. In order to avoid such an evolution toward a solid-body rotation, we should consider simultaneously the stellar and gaseous components of a galaxy. In such a case, the gas is driven by the stellar component. The gas viscosity then produces a spiraling motion of the gas toward the center. At the same time, part of the gas is transformed into stars; therefore, the gas becomes less dense and its effects less pronounced.

Our main conclusion from the preceding discussion is that any details smaller than the size h of the soft particles cannot be trusted. In order to improve our results we need an even smaller size h in the inner parts of the galaxy, but at the same time a larger number of particles in order to better represent the outer parts. We also need to consider a composite model consisting both of stars and gas, with a rate of conversion of gas into stars. Finally, one needs to take into account the cooling of the gas due to radiation. We are currently extending our calculations in these directions. Such calculations, even with improved SPH codes, however, need a very long time in order to provide reliable results concerning the evolution of a galaxy after the initial collapse.

REFERENCES

1. HAWLEY, D. L. & P. J. E. PEEBLES. 1975. Astron. J. **80**: 477.
2. SHARP, N. A., D. N. C. LIN & S. D. M. WHITE. 1979. Mon. Not. R. Astron. Soc. **187**: 287.
3. QUINN, P. J., J. K. SALMON & W. H. ZUREK. 1986. Nature **322**: 329.

4. BARNES, J. & G. EFSTATHIOU. 1987. Astrophys. J. **319:** 575.
5. ZUREK, W. H., P. J. QUINN & J. K. SALMON. 1988. Astrophys. J. **330:** 519.
6. VOGLIS, N. & N. HIOTELIS. 1989. Astron. Astrophys. **218:** 1.
7. MONAGHAN, J. J. & J. C. LATTANZIO. 1985. Astron. Astrophys. **149:** 135.
8. DURISEN, R. H., R. A. GINGOLD, J. E. TOHLINE & A. P. BOSS. 1986. Astrophys. J. **305:** 281.
9. HERNQUIST, L. & N. KATZ. 1989. Astrophys J., Suppl. Ser. **70:** 419.
10. LATTANZIO, J. C. & R. N. HENRIKSEN. 1988. Mon. Not. R. Astron. Soc. **232:** 565.

Stellar and Jovian Vortices[a]

T. E. DOWLING[b,d] AND E. A. SPIEGEL[c]

[b]*Astronomy Department*
Cornell University
Ithaca, New York 14853

[c]*Astronomy Department*
Columbia University
New York, New York 10027

> *Dond'escono quei vortici . . .?*
> —LORENZO DA PONTE
> *Don Giovanni*

1. INTRODUCTION

Large, long-lived vortices occur commonly throughout the Solar System. From the *Voyager* spacecraft, we have visual and thermal observations of the cloud-top regions of Jupiter, Saturn, Uranus, and Neptune. These reveal the existence of hundreds of atmospheric vortices on Jupiter, dozens of vortices on Saturn, and at least two large vortices on Neptune. Moreover, recent observations of the horizontal motions of solar granules provide convincing evidence for organized vertical vorticity on the Sun itself. To rationalize the common occurrence of vortices in the natural environment, numerical and laboratory experiments have recently been performed with rotating fluids. These demonstrate that long-lived vortices are suprisingly easy to produce, under a variety of environmental conditions.

This prevalence of vorticity distracts us from the surprising diversity of the observable motions in the Solar System's five rotating gas giants. Instead, our attention is focused on common features such as the resemblances between the Great Red Spot of Jupiter and the newly discovered Great Dark Spot of Neptune. We are prompted to ask whether great vortices may also be formed on rotating stars and what their existence might imply. We are led to imagine that analogous structures may exist on the most rapidly rotating stars, which are typically hot and subject to strong radiative forces. And we naturally wonder whether the impressive giant star spots seen on the rapidly rotating, solar-type, RS Canum Venaticorum stars may not be hydromagnetic analogues of the great Jovian vortices.

This paper is meant as an incitement to such thinking. We begin with an introduction to the problems of Jovian vortices, referring liberally to the extensive summaries of Jovian fluid dynamics already provided by several extant reviews (Ingersoll *et al.*;[1] Williams;[2] and Flasar[3]). In this, we adopt the current usage in

[a]This work has received financial support from the National Science Foundation under Grants PHY87-04250 and AFOSR89-0012 at Columbia University, and NASA under Grant NGL 33-010-186 at Cornell University.

[d]Current address: Earth, Atmospheric and Planetary Sciences Department, Massachusetts Institute of Technology, Cambridge, MA 02139.

190

planetary science of calling any vortex on a giant planet a "Jovian" vortex. Our discussion of these objects is made easier by the recent realization that the fluid dynamics of Jupiter's Great Red Spot may be understood without detailed knowledge of the underlying thermodynamics. This is in stark contrast to the study of hurricanes on earth, which we shall not describe here (but see Emanuel[4]).

The special characteristics of the Jovian problem are shared by many stellar envelopes. Jovian vortices are probably influenced by moist effects, like latent heat release and precipitation, whereas ionization plays a somewhat analogous role in the fluid dynamics of cool stars. But the "stellar" vortices may have their own complications. In the case of the cooler, solar-type stars, they occur in a background of intense, compressible, hydromagnetic turbulence, while in the hottest stars, radiation pressure is likely to be significant in the evolution of the vortex field. But the main difficulty in the study of stellar vortices is the sparcity of direct observation of the details of the fluid motions. That is why the observational study of stellar vortices is in its infancy, with the first solar vortex having been detected only recently[5] in the motions of of solar granules. The excitation mechanism is likely to be convective, as in the early simulations of Graham.[6] We shall refer to the recent theory of this process here.

In the case of the hottest stars, the observations are, at best, suggestive. The summary by Casinelli[7] raises the issue of solar-type activity on hot stars, including spots. We suggest that these spots need not be dark, like those on the sun; as we shall explain, both bright and dark spots may be expected. Nor are all spots necessarily produced magnetically. Rather they may be caused by vortices as in Bjerknes' now-forgotten ideas about sunspots.[8] We shall need to await simulations on the equations of photogasdynamics for detailed predictions on these questions. For, depending on the sizes and intensities of stellar vortices (that is, on their Rossby numbers), they may be like tornados or like Jovian vortices. Here, we can only offer some qualitative suggestions. But whether hot-star spots are vortical or magnetic, they arise in regions of anomalous gas pressure that will redirect the emergent radiation and produce surface inhomogeneities. In suitable conditions, vortices are conduits for escaping radiation and give rise to bright spots on a star's surface, as we shall discuss.

In many of the planetary situations that have been observed, there is a large, dominant vortex. This may also be the case in some stars, so it is important to know when this will happen. The reason for having a single dominant vortex may lie in the approximately two-dimensional character of rapidly rotating fluids. It is found that in predominantly two-dimensional rotating flow fields, as in atmospheric phenomena on large horizontal scales, smaller vortices tend to merge together into larger vortices. MacLow and Ingersoll[9] have reported dozens of vortex mergers seen in the *Voyager* images of Jupiter. Vortex mergers are commonly seen in laboratory and numerical experiments.[10-18] The formation of large vortices by the merging of smaller vortices is a manifestation of the fact that, in two-dimensional turbulence, the energy density flows toward larger scales, while the enstrophy density (squared vorticity) flows toward smaller scales.[13] Thus, one central issue when considering long-lived vortices on Jovian planets or stars is the source of smaller vortices.

The most important vortex source in the atmospheres of the Solar System's largest bodies is probably to be found in the strong surface winds possessed by all four giant planets and the Sun. These are all on the order of 100 m/s, measured relative to the rotation frames of their respective deep interiors. Yet these atmospheres have diverse dynamics, as we see from the comparison of their zonal wind profiles given in FIGURE 1. While looking at this figure, it is worth recalling that the Solar System's two giant vortices, Jupiter's Great Red Spot (GRS) and Neptune's Great Dark Spot (GDS), are both found at −23° latitude, and that both span about 20° in longitude and about 10° in latitude. It is also noteworthy that great vortices similar to the GRS and GDS are *not* found on Saturn, nor in the northern hemispheres of Jupiter and Neptune, nor in the observed (dynamic northern) hemisphere of Uranus, nor on the Sun. On the other hand, nearly every anticyclonic shear zone on Jupiter contains anticyclones, and nearly every cyclonic shear zone

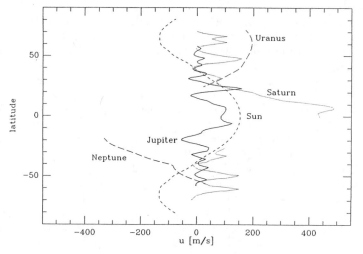

FIGURE 1. Observed surface zonal wind profiles for the Sun,[59] Jupiter,[60] Saturn,[1] Uranus,[61] and Neptune.[62] Velocity is defined relative to the rotating frame of the deep interior, which in the case of the Jovian planets is given by their respective magnetic field rotation rates, and in the case of the Sun is determined from helioseimisity.

contains cyclones. On all the giant planets, vortices are always found embedded in ambient shear zones (i.e., the vortices and their shear zones always have the same sign of vorticity). Thus, another central issue is the relation of zonal wind profiles to vortex formation. It will be interesting to see whether observations of solar vortices can bear on this point. More importantly, it should be asked whether analogous hydromagnetic processes play a role in forming the laws of sunspot polarity.

    In this discussion there are many such issues and it is hard to decide which are central and which may be omitted when we go to the stellar case. For example, Jupiter's GRS is an anticyclone: it rotates in the opposite sense to the planetary rotation, and hence is a high-pressure storm system. Now that Neptune's GDS appears to the *Voyager* imaging team also to circulate anticyclonically,[19] perhaps something ought to made of this clue. On Jupiter, about 85 percent of the long-lived vortices are anticyclones and about 15 percent are cyclones.[9] Nonlinear differences between cyclones and anticyclones, including both the balance between advection and dispersion, and differences involving moist convection, may explain why we find more anticyclones than cyclones in Jovian atmospheres.

In presenting here various aspects of vortices in the Solar System, we have clearly had to be selective. One question that warrants consideration is how a long-lived Jovian or stellar vortex can be maintained against dissipative losses and inviscid destructive processes. In Section 2, we begin our discussion with the basic problem of vortex maintenance. With this as background, we then examine in Section 3 what is known observationally about Jupiter's GRS, by far the most extensively studied Jovian vortex. This will point up the importance of zonal wind profiles in the Jovian vortex problem. In Section 4 we will enter briefly into observational and theoretical aspects of solar vortices. It seems likely that solar vortices arise by straightforward convective driving, a process that is also important on Jovian planets, with possibly a bit of help from solar rotation. By contrast, the violent activity on hot young stars, which are rapid rotators, may well arise in a large and active vorticity field driven by intense zonal winds. In Section 5, we speculate on this area of vortex dynamics, whose essential novelty lies in its two-fluid character, and whose basic conservation laws have yet to be exploited.

2. VORTEX MAINTENANCE

2.1. Introduction

Though Jovian vortices are subjected to several destructive processes, they survive for many turnaround times. Jupiter's Great Red Spot (GRS) has a turn-around time of six days, like that of many adjacent, short-lived atmospheric features, yet it has been observed for 300 years. The longevity of the GRS is perhaps less a mystery now than it was a decade ago, primarily because of recent successes in creating long-lived vortices in laboratory and numerical experiments. Still, we do not fully understand its remarkable persistence. According to Ingersoll,[20] the different experimental vortices are formed in a wide range of conditions that could not all be found in Jupiter's atmosphere. The models, however, collectively do have the value of helping to define a range of conditions under which stable vortices can exist. Thus, they lead us to anticipate the occurrence of vortices in similarly varied circumstances.

Theories of Jovian vortices are based on a diversity of approximations to the basic fluid equations. These approximations can be characterized by the classes of waves that are *retained*. Since the Jovian winds are subsonic, sound waves are taken out by the Boussinesq approximation, an approximation that is poor for certain aspects of the dynamics of stellar atmospheres. After sound waves, the next fastest moving waves are gravity waves, whose restoring force is differential buoyancy. Fluid equations that support these waves are called *primitive systems,* the simplest example being the shallow water equations. In Section 3 we review a new diagnostic shallow-water analysis of Jupiter's Great Red Spot.

The slowest atmospheric waves are called Rossby, or planetary, waves. The restoring force for these waves derives from a gradient in the ambient potential vorticity, a quantity we describe in the next paragraphs. Loosely, this is the ratio of absolute (inertial-frame) vorticity to layer thickness. The potential vorticity is of central importance in geophysical fluid dynamics because it is conserved by nondissipative fluid motions.[21,22] Equation systems that retain only Rossby waves are called geostrophic systems, with the particular class known as quasi-geostrophic systems[22] being the most tractable analytically.

2.2. Potential Vorticity

A potential vorticity, q, is formed by combining the conservation laws of mass and angular momentum for inviscid fluids.[23] This variable is central to the study of waves and the study of flow stability.[22] We assume viscous forces are negligible, and consider any scalar field λ that is conserved following the motion, that is

$$\left(\frac{\partial}{\partial t} + \mathbf{v} \cdot \boldsymbol{\nabla}\right)\lambda = 0. \tag{1}$$

The quantity

$$q \equiv \frac{(\boldsymbol{\nabla} \times \mathbf{v} + 2\Omega)}{\rho} \cdot \boldsymbol{\nabla}\lambda, \tag{2}$$

where $\mathbf{v}$ is the velocity, defined in a coordinate system rotating with angular velocity Ω, $\boldsymbol{\nabla} \times \mathbf{v}$ is the relative vorticity, and ρ is the density, is called a *potential vorticity*. If either (i) the fluid is barotropic ($\boldsymbol{\nabla}\rho \times \boldsymbol{\nabla}p = 0$), or (ii) λ is a function only of p and ρ, then the potential vorticity is conserved:

$$\left(\frac{\partial}{\partial t} + \mathbf{v} \cdot \boldsymbol{\nabla}\right)q = 0. \tag{3}$$

Since isentropic motion is a common first approximation, λ is often chosen to be the specific entropy. But any convenient variable that meets the preceding criteria may be used, as in Section 3.2.

Under rather general conditions, a necessary condition for flow instability is that there exist a local extremum in the unperturbed potential vorticity.[22] In Section 4 we outline one example of the application of potential vorticity conservation when we use it to infer the deep winds underneath Jupiter's Great Red Spot. We then characterize the stability of the cloud-top winds based on the resulting far-field potential vorticity profile.

2.3. Modeling Jovian Vortices

In modeling Jovian vortices, even the choice of the system of governing equations is a matter of dispute. The quasi-geostrophic equations produce a class of stable nonlinear solutions that are isolated vortices called *modons*.[24] The strong restriction on variations in layer thickness (or, equivalently, density variations) imposed in the standard derivation of quasi-geostrophic systems are, however, inappropriate to the Jovian conditions. As a two-layer analysis of *Voyager* velocity data shows,[25] the relative layer-thickness variations in the GRS are of order unity. Ameliorated systems of equations are now being fashioned. Gent and McWilliams,[26] have produced models intermediate between the standard quasi-geostrophic and the primitive equations. A development by G. G. Sutyrin of a general geostrophic two-layer system of equations that filters out gravity waves, and yet allows for order-one variations in layer thickness, is in preparation. We cannot go into the details of such models here, and we merely list them as the most promising starting point for a theory of stellar vortices. They should also be useful for the study of circulations in stellar atmospheres.

One of the problems in formulating basic approximations is that we know very little about the vertical structure of the Jovian atmospheres, even though we do have excellent flow visualization of Jupiter's GRS at the cloud-top level from the the *Voyager* images. The observed cloud features occur in a transition region between the nearly neutrally stratified, convecting deep interior, and the stable stratosphere. It is not known what the strength of stratification is around the cloud tops, but we can try to measure its global importance in terms of the Rossby radius of deformation, L_d. This is the distance over which the gravitational tendency to flatten fluid interfaces is balanced by the tendency of the Coriolis force to deform fluid interfaces.[22] To give immediacy to the meaning of L_d, it is helpful to relate it to specific models. Given the uncertainties of the problem, it seems best to develop models that do not rely sensitively on the details of the vertical structure, and the two-layer models fit this bill.

Consider a two-layer model with a constant upper-layer density, ρ_1, which is less than the lower-layer density, ρ_2. Then

$$L_d^2 \equiv \left(\frac{\rho_2 - \rho_1}{\rho_2}\right)\frac{gh}{f_0^2}, \tag{4}$$

where g is gravity, h is the upper-layer thickness, and f_0 is the typical value of the Coriolis parameter $f \equiv 2\Omega \sin (\text{latitude})$. The Rossby radius defines the scale above which rotational effects dominate those of stratification. It is useful to specify which regime one is in when dealing with phenomena of a given characteristic horizontal scale, L. This is done using a nondimensional stratification parameter called the Burger number,[22] S:

$$S \equiv \left(\frac{L_d}{L}\right)^2. \tag{5}$$

Of course, these quantities are constructed for plane-parallel fluid layers satisfying the Boussinesq approximation. The analogous quantities for the stellar cases are best defined otherwise.

Introduce the buoyancy (or Brunt) frequency, N, by $N^2 = g(\rho_2 - \rho_1)/(\rho_2 h)$. Then we have $S = (N/f_0)^2(d/L)^2$. But N can be given a local meaning without our having to specify a layer depth, so this formulation is better for continuous stratification, as in the stellar case, or even Jupiter. We can define an L_d with the local N in place of the global form, (4). Then we can use the condition $S \approx 1$ to define a depth d over which rotational rigidity (Taylor–Proudman effects) will overcome the effects of stratification. In stellar applications, this gives an expression for depth of penetration of convective motions of horizontal scale L into adjacent stable layers.[27]

It would be desirable to use such locally defined parameters for Jupiter, but we lack knowledge about the vertical variation of quantities such as the Burger number. On these grounds, models have been developed that drastically simplify the structure of the Jovian atmosphere. Most work on the Jovian vortices has been based on models with only one or two layers of constant density. What is surprising is how much such models teach us. In the anticipation that some of this work may prove interesting for the stellar case, we shall give a short catalogue of models of the Jovian atmosphere.

The simplest models are the one-layer models that confine the vortices to a single weather layer of constant density. These may be slightly extended to the so-called 1½-layer models, which allow for the effects of a steady zonal wind profile in a second fluid layer underneath the weather layer. Finally, we have the richer dynamics found in models with two or more fully active layers.

Two groups have recently presented results with one-layer laboratory experiments, using rotating water tanks. In order to minimize frictional (Ekman) boundary-layer effects the tanks must be rotated at high speed. Antipov et al.[28] (see also the review by Nezlin and Sutyrin[16]) studied a layer of water with a parabolic free surface. By using a parabolically shaped tank, they achieved a uniform depth of fluid in the undisturbed case. Once vortices were introduced, the layer thickness variations in them were of order 1, with a Burger number S typically less than one. The detailed effect of a vortex depends on whether its vorticity reinforces or opposes the planetary vorticity, that is, on whether it is cyclonic or anticyclonic. It is found that cyclones, which correspond to local free-surface depressions, last only about as long as the time for dispersive spreading of a linear Rossby wave packet, but that anticyclones, which correspond to local free-surface elevations, last much longer. We discuss this nonlinear asymmetry later. In any case, none of the vortices in this study survive, but succumb to eventual viscous decay.

In their experiments, Sommeria et al.[14] achieved vortex turnaround times an order of magnitude shorter than Ekman spin-down times by covering their large tank with a flat, rigid lid, and using very high rotation rates. An advantage of this technique is that their vortices persist, but the rigid upper boundary condition allows no layer-thickness variations. With a rigid lid, one has infinite Burger number. When $S = \infty$, all the vortices must be considered small and cyclones and anticyclones behave similarly. Atmospheric spots are found in the *Voyager* images with sizes ranging from the 20,000-km major diameter of the GRS, down to the 100-km or less scale. Thus, vortices found in both types of water tank experiments are relevant to the general Jovian problem. In particular, for the GRS, Marcus[15] argues that S for the GRS is of order one; he also believes (private communication) that, in any case, the value of S is not significant for the basic dynamics of Jovian vortices. The experiments will help decide these nagging issues.

On the theoretical side, Williams and Wilson[17] have presented an analysis of the stability and genesis of vortices in the one-layer, shallow-water analogue to the Jovian problem. They ran one isolated vortex in their numerical model for over 100 years. On examining the balances among the translation, twisting, steepening, dispersion, and advection processes, Williams and Wilson found that the advection term is the main preserver of vortices. In addition, they observed that anticyclones occurring near the equator, such as Jupiter's GRS and Neptune's GDS, can resist the highly dispersive equatorial modes to which they are subjected, if there is a strong equatorial westerly (eastward) jet and a significant easterly (westward) current. This is the case on Jupiter. Such an equatorial westerly jet probably does not exist on Neptune, however; features near the equator on Neptune are actually seen by the *Voyager* imaging team to keep up with the GDS. The problem might be resolved if an argument could be found that singled out the position of 23° in latitude, as well as the size of 20° in longitude by 10° in latitude.

Williams and Wilson's analysis was for a one-layer model with a flat bottom. It turns out that a two-layer model where the deep layer motions are prescribed to be steady and zonal is dynamically equivalent to a one-layer model with meridionally varying solid bottom topography, the so-called 1½-layer reduced-gravity model. The deep layer is meant to represent the neutrally stratified interior of the planet, and motions in this layer are assumed to be unaffected by the dynamics in the thin, overlying weather layer. Motions in the weather layer can, however, be affected by the deep motions through the variable bottom topography. With the vortex-tube stretching analysis described in Section 3, the *Voyager* GRS velocity data can be used to derive the appropriate bottom topography, and hence the deep motions, for such a model. Prior to this development the unwanted freedom to choose deep wind profiles existed, as in the next two models we discuss.

Theoretical considerations of neutrally stratified, rotating fluids motivated Ingersoll and Cuong[12] to think in terms of motions confined to differentially rotating cylinders coaxial with the rotation axis of the planet. They studied the case where the deep winds were identical to the far-field winds in the thin upper layer. For a quasi-geostrophic analysis, they showed that this deep wind assumption, together with a Burger number $S < 1$ ($k^{-2} < 1$ in their notation), meant that flow around a vortex would not generate Rossby waves, and hence that the vortex would be localized.

Rossby waves propagate along potential vorticity gradients, and a well-mixed system with uniform potential vorticity would produce no Rossby radiation. Such homogenization of potential vorticity is known, particularly inside closed streamlines.[29] It has also been encountered in laboratory experiments.[14] Yet it is not clear that the elimination of Rossby wave propagation in the far field is needed for Jovian vortices to isolate themselves. The Williams–Wilson[17] study, as well as earlier work they cite, shows that vortices with Burger number $S < 1$ can overcome the destructive effects of Rossby wave dispersion through a nonlinear balance with advection, as in ordinary solitary waves. But these are unlike solitons in that the equations have terms that cause interacting vortices to merge on close passage, instead of passing through each other. In any event, none of the known prescriptions for the deep motions in 1½-layer models are consistent with the GRS vorticity observations, as yet.

We now turn to models with two or more "vertical degrees of freedom." One complication is that systems with at least two layers (or continuous stratification) are subject to baroclinic instabilities,[22] that is, to instabilities resulting from interactions between different vertical modes.[30] The basic transfer of heat from equator to pole in the earth's atmosphere occurs at midlatitudes, primarily through developed baroclinic instabilities. We know too little about the vertical structure of Jovian atmospheres to say whether baroclinicity plays a decisive role here. Without trying to settle these issues, we simply mention some of the leading models.

Flierl[31] has studied Rossby wave radiation from a thin upper layer into a finite-depth lower layer. He finds that vortices initially constrained to the upper layer can suffer large energy losses by exciting Rossby waves on the fluid layer boundary. The rate of energy loss is proportional to the ratio of upper-layer to lower-layer thickness. For Jovian atmospheres, the latter ratio is essentially zero, so the energy loss should be negligible. Nevertheless, Flierl's work raises the possibility of strong qualitative differences in vortex evolution between 1½- and 2-layer models.

Read and Hide[32-34] have presented a series of experimental papers on isolated vortices found in rotating baroclinic systems. Stable vortices are seen to form under certain conditions involving internal heating or cooling of the fluid. Read and Hide show explicitly how their vortices maintain themselves against dissipation, via "slantwise" convection driven by horizontal temperature variations. An alternative hypothesis of Ingersoll and Cuong[12] is that large Jovian vortices maintain themselves against dissipation by absorbing smaller vortices, which are produced by convection. Both processes may in fact be important.

The models we have listed contain severally the essentials of vortex formation and maintainance. In geophysical fluid dynamics, they also form the backbone of theoretical discussion of circulation theory by means of which one attempts to understand planetary atmospheric dynamics and their simulations, both numerical and experimental. Now that acoustic sounding has revealed the rotational motions of the outer solar layers,[35] it becomes possible to adapt such models to stellar fluid dynamics. This prospect, in part, has motivated this discussion of selected models of Jovian fluid dynamics.

2.4. Cyclones and Anticyclones

We have used the words cyclone and anticyclone sparingly so far, and with little elaboration, so as not to slow the flow of the discussion. In fact, much of the recent work previously cited tells us that the physical natures of cyclones and anticyclones are surprisingly different. We need to describe the nature of these differences, which arise largely in the nonlinear balances in the two cases[17] and, to a lesser extent, in the action of moist convection.

The importance of nonlinearity in a rotating fluid is estimated by comparing the inertial force to the Coriolis force. The ratio is called the Rossby number, ϵ:

$$\epsilon \equiv \left(\frac{U}{f_0 L} \right), \tag{6}$$

where U is a typical speed. As before, L is a characteristic horizontal scale and f_0 is the Coriolis parameter. We concentrate here on rapidly rotating systems whose flows all have small Rossby number and are therefore geostrophic.

Williams and Wilson[17] analyze Jovian vortices using what they call the "geostrophic potential vorticity" equation. This equation is central to planetary vortex dynamics, and it involves all the parameters that we have already mentioned. It also contains another nondimensional quantity, the *beta parameter*, which measures latitudinal variations in the Coriolis effect:

$$\beta \equiv \left(\frac{L}{f_0}\right)\left(\frac{df}{dy}\right)_0.$$ (7)

It would not be helpful in this cursory review to display the equation derived by Williams and Wilson itself. The key fact for our present purposes is that there is such a nonlinear differential equation central to the dynamics and that it contains a number of important parameters whose values are not known very well.

In the circumstances, we can only outline the key parameter ranges with their characteristic behaviors, as a guide to discussion of the stellar possibilities. Three regimes for the Burger number that have been examined in some detail are the "quasi-geostrophic" regime ($S \sim 1$), the "intermediate-geostrophic" regime ($S \sim \epsilon^{1/2}$), and the "planetary-geostrophic" regime ($S \sim \epsilon$), with $\beta \sim 1$. In the case $S \sim 1$, the nonlinear steepening terms that appear in the main equation drop out. Thus cyclones and anticyclones, which correspond to negative and positive pressure anomalies, respectively, behave similarly in this regime. For small S, these nonlinear terms are important, and the sign of the anomaly is crucial to the nonlinear balance. Analysis of the geostrophic potential vorticity equation shows that anticyclones (with locally raised surface elevations) can achieve a balance between the dispersive and steepening terms, whereas cyclones cannot.

As already mentioned, we do not know what value to assign the Burger number in the Jovian atmosphere. Typical estimates of the Rossby radius are $L_d \sim 1000$ km. Care must be taken when deciding on the length scale L. If one takes L to be the semimajor axis of the GRS, then one finds $S \sim 0.01$. As pointed out by Marcus,[15] however, a more relevant scale is the distance over which the velocity of the vortex varies, which he estimates to be $L \sim 2000$ km, yielding $S \sim 0.25$. It appears that for the GRS, S is less than one, but ignorance of the value of L_d makes this at best an uncertain conclusion.

Even more complicated issues arise when we try to understand the role of convective dynamics in these problems. There is a strong correlation between cyclonic regions, convective activity, and folded, filamentary cloud morphology on Jupiter and Saturn. Three well-documented examples are the convection in the cyclonic, filamentary region west of Jupiter's GRS,[36] the transition of an oval-shaped cyclone into a folded, filamentary cyclone on Jupiter,[36] and the triggering of increased activity of a convective region on Saturn by the nearby passage of a cyclone.[37] Attempts to explain such correlations involve us in subtle issues of "moist convection." The pursuit of such details again reveals distinctions between the physics of cyclones and anticyclones.[38] Here we simply want to indicate the existances of such nuances for the sake of those intending to pursue similar issues in the stellar case.

With this glimpse of difficulties at the frontiers of Jovian fluid dynamics, we leave the problem of maintaining vortices and turn to an observational aspect of Jovian studies. This involves a close look at the *Voyager* data of Jupiter's GRS. We hope that the approach described may ultimately be of value in the diagnostics of solar gyromagnetics, but we also appreciate that some of the features of the discussion are technical for readers not fluent with the language of geophysical fluid dynamics. Since the sections on stellar vortices do not refer to the discussion in Section 3, however, it is possible to pass on to those directly without losing the thread of the discussion.

3. JUPITER'S GREAT RED SPOT

3.1. Introduction

In this section, we describe how data on cloud velocities have been used to probe the vertical structure of the Jovian atmosphere so as to provide some of the needed ingredients of simple models for the GRS. The high-resolution data come from only a few brief flybys, and we might therefore worry about the fact that we derive only a kinematic snapshot of velocity fields that may be in continual transformation. However, there are some comforting observations on this score. Jupiter's atmosphere has been observed quite steadily over the last century, and it has varied greatly in cloud activity, color, and contrast.[39] The GRS was once substantially longer in longitudinal extent.[40] Also, at the time of the 1979 *Voyager* encounters, strong color contrasts and heavy activity occurred. Though it had been quiescent five years earlier at the time of the *Pioneer* spacecraft encounters, the large convective filamentary region to the northwest of the GRS was active. The two *Voyager* spacecraft encounters were separated by four months, and while they found virtually no change in the cloud-top velocity field, they did reveal some changes in cloud color and appearance. This may mean that the sources of cloud color vary more rapidly than does the velocity field, so the results presented in Section 3.2 may be robust.

The velocity data from Jupiter appear also to be of good quality. The *Voyager* time-sequenced images provide an excellent flow visualization of Jupiter's GRS and other ovals, one that is substantially better than the typical flow visualization available for the earth's Gulf Stream, for example. Velocity is determined by tracking small cloud features in images taken typically ten hours (one rotation period) apart. The resulting data for the GRS are reproduced from Dowling and Ingersoll[25] in FIGURE 2(a), with computed streamlines shown in FIGURE 2(b). We can look forward to similar data from the Hubble Space Telescope and the Galileo spacecraft observations, to ascertain the extent of changes that have occurred over the last decade.

3.2. Data Analysis

For the diagnostic analysis of the velocity data in FIGURE 2, we use a 1½-layer shallow-water model.[18] To clarify this, we first give the so-called shallow-water equations, which are the equations of conservation of momentum and mass. They govern the horizontal velocity field $\mathbf{v}$ with eastward and northward components u and v, and the layer thickness, h. The inviscid momentum equation may be written:

$$\mathbf{v}_t + (\zeta + f)\mathbf{k} \times \mathbf{v} = -\nabla B, \tag{8}$$

where $\mathbf{k}$ is a vertical unit vector, $\zeta \equiv \mathbf{k} \cdot (\nabla \times \mathbf{v})$ is the vertical component of relative vorticity, $B \equiv [g(h + h_2) + K]$ is the Bernoulli function, in which $K \equiv \frac{1}{2}|\mathbf{v}|^2$ is the kinetic energy per unit mass.

We assume that the fluid is incompressible, so that the continuity equation is an equation for the layer thickness:

$$h_t + \nabla \cdot (h\mathbf{v}) = 0. \tag{9}$$

The variable gh_2 in the Bernoulli function is called the *bottom topography*, where g is taken to be the reduced gravity, and it represents the horizontally varying part of the deep-layer pressure field, divided by the density ρ_2. It is constrained, such that:

$$fu_2 = -(gh_2)_y. \tag{10}$$

As we mentioned earlier, we have no direct observations on the deep wind, u_2,

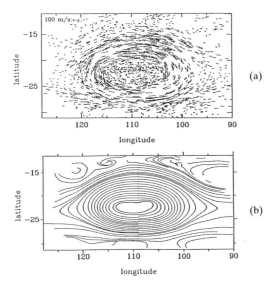

FIGURE 2. (a) The *Voyager* GRS cloud-tracking velocity data. The locations of the vectors are marked with *dots*, and the *lines* point downwind. (From Dowling and Ingersoll.[25] Reproduced by permission.) (b) Streamlines computed by integrating the velocity data in (a). *Small dots* indicate intervals of 10 hours.

which, until recently, has had to be given arbitrary specifications. But it may in fact be determined from the observations with the help of these equations.

Since each layer of a shallow-water model has constant density, the flow inside a layer is barotropic ($\nabla\rho \times \nabla p = 0$). In shallow water, w is linear in z, and so the quantity called the status function of a fluid element, $(\rho/g)(z/h)$, is conserved[22] and can be used to form a potential vorticity:

$$q \equiv \left(\frac{\zeta + f}{gh} \right). \tag{11}$$

The conservation of q in a shallow layer of incompressible fluid means that total vorticity is altered by the stretching of vortex filaments by changing layer thickness. It

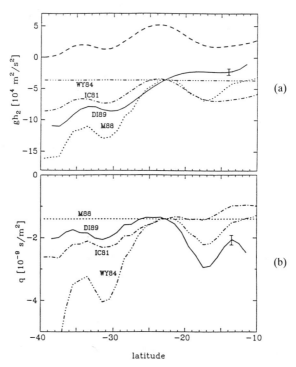

FIGURE 3. Results of the shallow-water analysis of Dowling and Ingersoll.[18] (a) The *dashed line* shows the free-surface height, $g(h + h_2)$, determined by integrating the zonally averaged cloud-top velocities. The profile marked DI89 is the bottom topography deduced from the analysis, for the case $q_0 = -1.4 \times 10^{-9}$ sm^{-2}. The *error bar* shows the deviation of the data about the model fit, and is two standard deviations in total length. The profiles marked IC81, WY84, and M88 are the bottom topographies prescribed in the models by Ingersoll and Cuong,[12] Williams and Yamagata,[41] and Marcus,[15] respectively. (b) Corresponding potential vorticity profiles.

can also be seen how this quantity might relate to the local elevation of the free surface.

In a reference frame drifting with the GRS, the motions in and close to the vortex are approximately steady, and the Bernoulli function B is also approximately conserved. This is seen on dotting (8) with **v**. In addition, B may be computed from the velocity data by integrating (8).

FIGURE 4. Vortex genesis experiment from Dowling and Ingersoll.[18] The initial state is constructed by adding a small sinusoidal perturbation of ten wavelengths in longitude to the observed zonally averaged cloud-top wind profile of Jupiter, together with the derived bottom topography shown in FIGURE 3(a). The zonal wind is forced on a 400-day timescale. The peak initial wind is 54 m/s, and the peak perturbation velocity is 1.6 m/s. The system is unstable, and by $t = 500$ days three large, distinct vortices have formed in the GRS's shear zone. By $t = 750$ days, two of these vortices have merged, and by $t = 1600$ days, only one large vortex remains. (From Dowling and Ingersoll.[18] Reproduced by permission.)

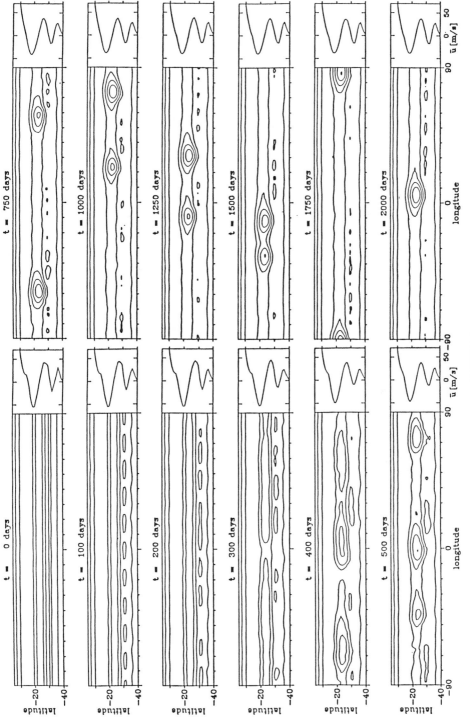

FIGURE 4

Using the preceding relations, we may express the bottom topography gh_2 as

$$gh_2 = B - K - \frac{\zeta + f}{q}, \qquad (12)$$

where the only unknown on the right-hand side of (12) is q. Since both q and B are conserved following the motion, we may label streamlines with B, and write $q = q(B)$, taking care to allow $q(B)$ to be multivalued when different streamlines have the same value of B. We proceed by specifying the value of q on a particular streamline labeled by B_0, that is, $q_0 = q(B_0)$. We model gh_2 as a quartic in latitude, y, so that the profile of the deep wind, u_2, may resemble a cubic, and model $q(B)$ as piecewise quadratic in B. The parameter q_0 is our only free parameter. A least-squares fit to the velocity data in the form (12) simultaneously yields $q(B)$ and $gh_2(y)$.

In FIGURE 3(a) we reproduce the derived gh_2 determined from the GRS data, with an extention to higher latitudes using data from the White Oval BC centered at $-33°$ latitude. The results are for a typical value of the free parameter q_0, and are presented together with the prescribed gh_2 profiles from the models of Ingersoll and Cuong,[12] Williams and Yamagata,[41] and Marcus.[15] The corresponding potential vorticity profiles are shown in FIGURE 3(b).

Although the prescribed gh_2 and q profiles match the derived profiles over part of the range of latitude, none of the prescribed profiles agrees with the derived profiles over the whole range, given any value of q_0. This means that the current theories are too simple, and that we must now come up with an improved, simple characterization of the deep winds.

The derived profile does, however, meet the necessary condition for instability in exhibiting a local extremum in the potential vorticity profile. We can test the stability of the system in a numerical initial-value problem. FIGURE 4 shows the result of such an experiment, where the upper layer is initialized with Jupiter's zonally averaged cloud-top velocity profile, plus a small sinusoidal perturbation, together with the derived bottom topography shown in FIGURE 3(a). The anticyclonic shear zone at $-23°$ is seen to be unstable, and breaks into a number of small anticyclones. These smaller vortices coalesce into a single large vortex on a timescale of the order of a decade. If the initial wind profile is forced, say, on a timescale of 400 days, then the instability is maintained and smaller vortices are continuously produced and swept up into the larger vortex.

This analysis suggests that the vortices on a Jovian planet are simply due to a maintained unstable zonal wind profile, and represent the fully nonlinear evolution of such a system. If this is indeed the case, then we ought to turn our attention to the nature of zonal winds on Jovian planets. What are the sources of the shear? Or, more technically, what are the sources of potential vorticity extrema, and how is the tendency for potential vorticity toward homogenization overcome? If we could come up with a way to maintain unstable zonal winds, we could produce an atmosphere covered with large, long-lived vortices. On the other hand, we must also consider the possibility that such vortices have something to do with producing the conditions that sustain them. In that case, the possibility of long-lived vortices bootstrapping themselves into favorable conditions may be what we need to understand.

4. VORTICES ON COOL STARS

4.1. Introduction

White light photographs of the solar surface reveal a mottling on a scale of the order of 2000 km known as solar granulation. These bright objects are generally regarded as rising convective parcels, with possibly complicated internal dynamics. Granules have rise velocities of perhaps half a kilometer per second, for a Mach number of a few tenths. On a scale ten times larger, one sees another apparent manifestation of the solar convective motion called *supergranulation.* Supergranules are detected kinematically as horizontally diverging from reasonably distinct centers at about 1 km/s. Supergranules resemble the tops of convective cells in their motion. The supergranules are outlined by relatively strong concentrations of magnetic field of up to about 2000 G. It also appears that the granules organize themselves in collective behavior on a scale of about five granule diameters; this is called the *mesogranular scale.* Simon and Weiss[42] have given a brief summary of the basic papers on these matters.

Solar convection is too complex for us to attempt a theoretical review here. Suffice it to report that mixing length calculations have been used to get some idea of the structure of the outer convection zone of the sun. This layer goes about a third of the way in to the center of the sun and it is effectively homentropic, a partly theoretical result that may be confirmed by the study of solar oscillations.[43] It seems that for many purposes, a two-layer model for large-scale solar motions might be suitable, with the outer layer, the solar photosphere, being the weather layer in which we see the granulation.

The buoyancy frequency in the photosphere, N, is about 0.01 s^{-1}. Hence $L_d \approx$ 200d, where d is the layer thickness. If we assume that the mixing-length models give us the depth of the weather layer correctly, we have $d \approx 10^3$ km and $L_d \approx 10^5$ km. So $S \gg 1$ for the granulation and ≈ 0.2 for the supergranulation. Beyond this, Simon and Weiss suggested that there is also a large-scale regularity on the scale of 10^5 km that led to the possibility of a large scale of motion, the so-called giant cells. Mullan suggests, however, that such a large-scale order ought to be a result of vortices on this scale rather than convection. Nor was this the first discussion of the possibility of solar vorticity.

Bjerknes[8] suggested that sunspots are caused by vortex tubes that bow up and pierce the solar surface in a pair of spots, much as magnetic-flux tubes are thought to do in modern theories. Though discredited as an explanation of sunspots,[44] vortices may play a role in spot production on other stars, as we shall see in the next section. Moreover, Akosoka[45] has pointed to the appearance of a particular spot group on the sun that is suggestive of a strong solar vortex. Whatever the case in these beginnings, there is now kinematical evidence for vertical vorticity on the solar surface.

Observations of high resolution of the solar surface have led to the detection of solar vortices.[5] This discovery relies on the recently developed technique of filtering of acoustic oscillations from the observations to make it possible to follow the motions of individual granules. Without this filtering, a typical granule had an apparent lifetime of 5 or 10 minutes, just long enough to travel its own diameter.

Now, granules can be followed for longer than their turnover times, and this has advanced the study of granular motions and solar surface motions in general.

The horizontal motion of granules reveals the vertical component of the vorticity field. This is the quantity of principal interest in the study of solar vortices, since the horizontal vorticity component is associated with the ordinary convective motions. That is, the horizontal and vertical vorticity components are associated with distinct modes of the linear theory of convection,[46] though these modes become mixed in the case of a rotating fluid.

The generation of vertical vorticity by convective motions is a natural outcome of nonlinear mode coupling, but the strength of the effect was surprising when it was first seen in three-dimensional simulations of compressible convection by Graham.[6] The lively vortical motion produced in Graham's simulations can be understood as a combination of two processes. First, there is the growth of the vertical vorticity as a secondary instability as we shall describe next. Then there is the process of vorticity concentration by turbulence, which is not well understood, and about which we shall remain reticent, except to refer to the experimental work of McEwan[47] and Hopfinger and Browand.[48] The process they describe resembles that thought to generate zonal winds on rotating planets.

4.2. Convective Generation of Vertical Vorticity

The estimate of a Rayleigh number for the solar convection zone is difficult, and not just because its density varies over six or more orders of magnitude. The Rayleigh number is defined from the parameters of the conductive (here radiative) state, which people do not consider worth calculating. Estimates may therefore vary, but it seems that the solar Rayleigh number is about 10^{20}, if not more. This implies an intense level of turbulence, and it may seem surprising that one sees regular structures like granules or vortices.

Given the present poor understanding of turbulence, it may be out of place to express surprise at any of its manifestations, but the detection of ordered structures in intense turbulence does seem to call for rationalization. The usual reason given for the semblance of order and of coherent structures in intensely turbulent flows is that the turbulence somehow renormalizes itself to near neutrality. In response to the enhancement of heat flux by convection, the temperature gradient's absolute value diminishes so that it is nearly the same as the adiabatic gradient. This is the state called *convective equilibrium* by astrophysicists since the late nineteenth century.

Moreover, the turbulence that does occur also causes an increase of the effective diffusivities. Under these influences of the turbulence, it is possible to estimate an effective Rayleigh number that is quite close to the critical value for the onset of convection.[49] In that context, one may risk discussing ordered structures as if the situation were only mildly unstable and still hope to draw some useful qualitative conclusions. Our discussion of the origin of solar vortices is founded on this vision, but there is another way to understand the prevalence of ordered structures in turbulent flows that may be helpful in thinking about this complex problem.

In the 1950s and the decades just before, there was an interesting qualitative vision abroad of what turbulence is that was mainly attributed to Hopf,[50] though it is hard to extricate specifically from his writings. The picture is that there is a large (probably infinite) number of simple or laminar solutions of the equations of fluid dynamics that are all unstable in conditions of fully developed turbulence. The full solution in such a case wanders about in the phase space from one to another of these special solutions. The representative point in phase space will naturally move quite slowly as it goes by those of the ordered solutions that are stationary or slowly varying, so that these will dominate the temporal averages and be quite visible to the casual observerer. A simple realization of this vision is offered by systems of ordinary differential equations that exhibit chaos.[51]

Turbulence is full of ordered structures, whose description has become better over the years. The simplest version begins in the linear theory of convection, where a number of normal modes are found that can be used to interpret and describe behavior. Besides the familiar thermoconvective modes that go unstable, there are the modes of vertical vorticity that we have mentioned. These have purely horizontal motion involving no thermal perturbations.[46] These vertical vorticity modes are purely damped on the timescale of $(\nu k^2)^{-1}$, where ν is the kinematic viscosity (here assumed to have its turbulent value) and k is the horizontal wave number. In the development of nonlinear convection, we expect such modes to be excited by nonlinear couplings to the convective modes, and so get their fair share of the kinetic energy. Their strong excitation in Graham's simulations, however, exceeded qualitative expectations. When they also started to loom large in laboratory experiments, it seemed likely that something else was happening.

Busse and collaborators[52] realized that convection is subject to a secondary instability in which steady two-dimensional convection becomes unstable to perturbations by vertical vorticity modes. A convective motion will bend a vertical vortex so that it has a slight component of vertical velocity, so the secondary instability is not so surprising with hindsight. Perhaps the best way to use such hindsight is to look back at the work of Chandrasekhar[53] on the instability of rotating convection. In that problem, vertical vorticity modes are linearly coupled to the convective modes through the rotation, so they are already implicated in the original instability leading to convection. If there is no rotation originally, we can, in nonlinear instability theory, imagine that a large-scale vertical vorticity mode is an allowed nonlinear perturbation, and this acts rather like a uniform rotation. In other words, the free vertical vorticity mode can be coupled to convection in either way. Given that the Burger number is of order unity for the observed vortices, it is likely that both processes are involved in their dynamics.

In nonlinear stability theory, the dynamics is generally controlled by the slow modes. Weakly nonlinear convection, in small containers, is controlled by the nearly marginal modes because the fast modes come quickly into some kind of equilibrium. In highly unstable cases, we suppose that very unstable modes generate turbulence, which renormalizes the transport coefficients to such large values that no modes are highly unstable for long. Once again, the slow modes will dominate in the mean. In the context of weakly nonlinear theory, the slow modes are the analogues of the ordered structures of turbulence and their dominance arises for similar reasons.

In a large container, the vertical vorticity modes with small horizontal wave number will also be slow and must be included in even the simplest descriptions of the convective process. Zippelius and Siggia[54] have shown the way to do this in the simplest examples. As they are slowly evolving, the large-scale vortices may be expected to play an important role in the dynamics. So their appearance in the sun is encouraging, even though they seem rather more weakly represented than one might have hoped. Perhaps they suffer more from turbulent viscosity than we have realized.

The theory that has been developed for nonlinear vortical motions has so far concentrated on the Boussinesq case, where the underlying convection is essentially two-dimensional. Even that case becomes richly complicated when there is an imposed rotation. But in the sun, the convective pattern is more complicated than in the typical laboratory case, and a theory has yet to be developed. The modal approach is one tool for the study of the nonlinear dynamics of vertical vorticity that may prove useful in going to highly unstable regimes where the intensity of the vertical vorticity may provide a diagnostic tool for turbulent viscosity.[55]

4.3. Prospects

In the next few years, we can hope to have a fuller observational picture of the vorticity on the solar surface. Given what we already know, it is natural to anticipate that there will be many vortices on the surface of the sun. But what determines their sizes? It may just be that the sizes of the first few vortices detected have something to do with the method of detection and that there really are a lot of vortices on larger scales. After all, the observed vortices are comparable in size to mesogranules,[56] only a few percent of the Rossby radius, so they are getting no help from the global solar rotation. On the other hand, there has been a suspicion that mesogranules are clumps of granules behaving in some collective way. Perhaps this collective behavior is caused by the vertical vortices themselves in a cooperative process.

Another line of inquiry ought to be directed at solar-type stars that rotate much faster than the sun, the RS Can Ven stars.[57] It is suspected that N (the buoyancy frequency) is not much different for these stars than for the sun. Hence, they have much smaller Rossby radii. Powerful rotational effects are to be expected on even the mesogranular scale on such stars. We can expect them to display intense vortices, but whether we should expect a single, dominant one like the GRS, or many competing vortices, we cannot yet say. The difference from the planetary case caused by the hydromagnetic aspects of the problem are too great. But that difference is the most interesting aspect of the surface dynamics of the cooler stars.

As far as kinematics is concerned, the magnetic field and the vorticity satisfy the same advective–diffusion equation. In the case of the magnetic field, this equation is linear, and it is the starting point of kinematic dynamo theory. There is no analogous theory of vorticity, since the connection between velocity and vorticity makes this equation strictly nonlinear. But perhaps this is not the main difference between the magnetic and vortical problems since, after all, both problems are strongly nonlinear in their full realizations. The reason that one does not treat the two problems along more similar lines probably lies in the fact that there is a reservoir of vorticity in the underlying rotation of the parent body, but no obvious analogue in the magnetic case. Some of the observational issues suggest, however, a reexamination of the question.

Though spots on the sun are loosely confined to a particular latitude zone at any time, they are rather widespread over the solar surface. On the other hand, the RS Canum Venaticorum (RS Can Ven) stars each appear to have a single strong spot, as far as can be told at this remove. Physically, the kinds of stars are quite similar, the main difference being that the RS Can Ven stars are rapid rotators. Perhaps the difference between the two kinds of stars is rather like that between terrestrial and Jovian planets, with the single spot in the RS case being hydromagnetic analogues of the GRS. Simulations like those done for the GRS would be enlightening. Instead of trying to understand the full dynamo problem, it might be of value to simply postulate a strong underlying toroidal field—the analogue of the basic rotation—and see what kind of spot field develops as a function of rotation rate. In this way, one might hope to bring together the results of planetary and stellar atmospheric dynamics. Simplified models like those reviewed in Section 2 will also prove enlightening in these problems.

5. SPOTS ON HOT STARS

The problem of solar hydromagnetics offers no evident simplifications and, although the case of the RS Can Ven stars may be more hopeful, the turbulence is still an obstacle to definite progress. The prospects for understanding the hot stars seem better because their atmospheric convection is relatively mild and most hot stars rotate rapidly. Yet the hot stars too have complicated fluid dynamics, and perhaps they are also turbulent. Our optimism may be unfounded, and without high-resolution observations, we can only speculate about what is happening in hot stellar atmospheres. In this section we offer a qualitative vision of the fluid dynamics of these objects.

Struve and Huang suggested long ago that the large radiation pressure in the hot stars plays a significant role in their dynamics.[58] Accepting their premise, we are interested in rationalizing some of the salient features of the observations[7]—large linewidths, variability, and strong winds. To do this, we must also allow for another significant feature of many hot stars: they are generally rapid rotators. For this reason, we are inspired by what we see on the Jovian planets. Rapid global rotation suggests large differential rotation.

Though we cannot reliably predict the details of the differential motion, we can estimate the magnitude of the velocity differences to be expected in the stellar surfaces, assuming that the differential rotation is driven by the rotational flattening. Rotational flattening is something that Newton knew how to estimate and is of the order of

$$\frac{\Delta R}{R} \approx \frac{\Omega^2}{\Omega^2 + g\rho} , \qquad (13)$$

where $\Delta R = R_{equator} - R_{pole}$ and Ω is the stellar angular speed. This produces a pole-to-equator temperature difference of about

$$\frac{\Delta T}{T} \approx \frac{\Delta R}{R} \qquad (14)$$

and a corresponding pressure difference

$$\Delta p \approx \Re \rho \Delta T \approx \Re \rho \, T \frac{\Delta R}{R} \approx p \, \frac{\Delta R}{R} . \tag{15}$$

The pole-to-equator pressure gradient is approximately balanced by the Coriolis force:

$$\Delta p \approx \rho u R \Omega, \tag{16}$$

where u is the *zonal* velocity. The vertical motion is relatively weak since the vertical stratification is strong and stable. Hence, on the short rotational timescales, the vertical forces are predominantly hydrostatic, so

$$\frac{p}{R} \approx g\rho, \tag{17}$$

where g is the atmospheric value of the gravitational acceleration. Putting all this together, we get the thermal wind relation, in qualitative form:

$$u \approx \frac{g\Delta R}{R\Omega} . \tag{18}$$

Many hot young stars rotate so fast that their equatorial centrifugal accelerations $R\Omega^2$ are comparable to g. Hence, it is normal to find $g/\Omega \approx R\Omega = V_{eq}$. We see that $u = V_{eq}\Delta R/R$; hence, the differential motion is comparable to the equatorial velocities. Whether we then use atomic or radiative viscosities, we find very large Reynolds numbers for the atmospheres of the hot young stars. This simple picture of the differential rotation also indicates that there will be at least one jet in the zonal flow, presumably equatorial, with attendant inflection points in the velocity profile. In short, all the evidence points to the generation of turbulence and of strong vortices in the atmospheres of young main sequence stars. This could well be connected with the large linewidths in such stars that Struve repeatedly emphasized and attributed to *macroturbulence*.

The most striking consequence of the prevalence of vortices over the surfaces of hot stars will result from their intense radiation fields. For a qualitative impression of these effects we can make a simple vortex model. To avoid the question of whether we are dealing with cyclonic or anticyclonic effects, let us assume an intense vortex in which the stellar rotation is not locally a significant factor. Such a model will not be useful in discussing the motions of the vortices or their temporal evolution, but it gives us a first impression of what consequences may come from creating a vortex in a hot stellar atmosphere.

Consider, then, a vortex in a polytropic atmosphere. Let the vorticity be purely vertical with circular velocity v around its axis. Our picture is that the vortices are being continually generated by instability and damped by viscosity, perhaps even keeping alive by the sort of cannibalism seen on Jupiter. Here we maintain stationarity by leaving such terms out.

The mean properties of our simple vortex, in cylindrical coordinates, with the vertical coordinate, z, taken positively downward, will be given by these balance equations:

$$\frac{dp}{dz} = g\rho \tag{19}$$

$$\frac{dp}{dr} = \rho\,\frac{v^2}{r}\,. \tag{20}$$

For a polytropic medium, $dh = dp/\rho$, where h is the specific enthalpy. From (19) we learn that

$$h(r,\, z) = gz + f(r), \tag{21}$$

where f is an arbitrary function. According to (20) f satisfies

$$f' = \frac{v^2}{r}\,. \tag{22}$$

The latter result implies that v is independent of z.

We adopt the standard potential vortex, with $v = v_0 r/r_0$ for $r/r_0 < 1$ and $v = v_0 r_0/r$ for $r/r_0 > 1$, where r_0 and v_0 are constants. For $z = 0$, we require the surface to be flat at $r \to \infty$. Then we get

$$f = v_0^2 \begin{cases} \dfrac{r^2}{2r_0^2} - 1, & \text{if } r < r_0; \\[2ex] -\dfrac{r_0^2}{2r^2}, & \text{if } r > r_0. \end{cases} \tag{23}$$

The vortex deforms the atmosphere's surface from the plane $z = 0$ into

$$\frac{gz}{v_0^2} = \begin{cases} 1 - \dfrac{r^2}{2r_0^2}, & \text{if } r < r_0; \\[2ex] \dfrac{r_0^2}{2r^2}, & \text{if } r > r_0. \end{cases} \tag{24}$$

So the vortex represents a pit in the stellar atmosphere with a central depth $z_0 = z(r = 0) = v_0^2/g$. The contours of the thermodynamic quantities follow that of the surface, but are vertically displaced.

How are these formulas modified by radiation pressure? That is a question of ongoing research, but there are circumstances in which there is no qualitative change in this picture. For a polytropic atmosphere with $p = K\rho^\Gamma$, we have $h = \Gamma K/(\Gamma - 1)\rho^{\Gamma-1}$. If the medium is also a perfect gas with $p = \Re\rho T$, we get $h = c_p T$. Now the radiation pressure is $p_{rad} = \frac{1}{3}aT^4$, so the total pressure is $P = p + p_{rad} = p(1 + \beta)$, where β is a constant times $\rho^{3\Gamma-4}$. For the plausible condition (for hot stars), $\Gamma \approx 4/3$, the dependence of pressure on temperature is virtually unchanged and the vortex structure survives as described.

The main observational consequence of having a strong, deep vortex in the atmosphere is the attendant modification of the radiation field. The qualitative idea is illustrated by the use of hydrogen bubbles as markers of fluid motion.[48] When small bubbles are introduced in the bottom of a tank of water with strong vertical vortices, the bubbles move into the vortices. One clearly sees the bubbles accumulating in the vortex cores, there to move upward along the path of smallest resistance. This behavior will be imitated by any lighter fluid introduced in the main fluid.

To see how this flow into the vortices operates for the radiation fluid, let $\mathbf{F}$ and E be the radiative energy flux and energy density, respectively. In the case of a stationary vortex, in the diffusive limit, we have

$$\mathbf{F} = -\frac{c}{3(k + \sigma)\rho}\,\nabla E \tag{25}$$

and

$$\nabla \cdot \mathbf{F} = \rho k c \left(\frac{4\pi}{c} B - E\right), \tag{26}$$

where k and σ are the absorption and scattering coefficients of the medium, $B = (\sigma_*/\pi)T^4$ is the integrated Planck function with σ_* the S–B constant, and c is the speed of light. (If it is desired to allow for the Compton effect in this description, it is best to include it in the absorption coefficient.) In a hot star, we normally have $k/\sigma \ll 1$, so we might expect any deviation from radiative equilibrium to last for some time. But for this qualitative exploration we use the equilibrium condition $E = 4\pi B/c$, to reduce the transfer problem to this simple form:

$$\mathbf{F} = -\frac{16\sigma_* T^3}{3\sigma\rho}\,\nabla T. \tag{27}$$

For a polytrope, the contours of ρ and of T are coincident; for $\Gamma = 3/4$, T^3/ρ is a constant. In that case, we have the simple approximation that the radiation flows down the temperature gradient. Since $h = c_p T$, **(21)** and **(23)** give, for $r < r_0$, the isotherms

$$T = \frac{g}{c_p}z + \frac{v_0^2}{c_p}\left(\frac{r^2}{2r_0^2} - 1\right). \tag{28}$$

Since T is approximately the potential for $\mathbf{F}$, we see that radiation flows toward the vortex core and continues on into the surface depression made by the vortex, as illustrated in FIGURE 5. A strong vortex will produce a deep hole, which will serve as a sort of light well. Photons gushing from it will sweep material from the vortex core along with it. This motion will seed a stellar wind.

We have not resolved the quantitative problems involved in this process. The local emerging radiative flux is enhanced over its undisturbed value by a factor of about $(z_0/r_0)^2$, that is, about $[v_0^2/(gr_0)]^2$. This is the quantity that the theory has to produce, but it is not a quantity that can be reliably estimated until we know whether we are dealing with a single giant vortex or many small ones. We have illustrated the focusing effect on outgoing radiation due to the low-temperature cores associated with small, intense cyclones. This produces a bright spot on the stellar surface. If the analogy to the planetary case works, such spots should be short-lived. The analogy also suggests that there may be large long-lasting anticyclones. Though any focusing or (more likely) defocusing of radiation by giant vortices will probably be relatively mild, since the analogy suggests that their horizontal extents far exceed their depths, their persistence may make them detectable. Perhaps too giant vortices will produce long-lived magnetic anomalies that will make them observable, much as Jupiter's Great Red Spot is observable by the reddish clouds trapped inside it. What is needed is a fast UV photometer such as Bless has devised for the Space Telescope.

But details aside, it appears that rotational instabilities shaped by radiative forces can be invoked to rationalize some of the salient features of the observations of hot stars such as large differential velocities, generation of coronas, and the start-up of winds. It is therefore worthwhile to begin thinking seriously about rotating radiative dynamics. This problem, central to astrophysics, has been remarkably neglected so far. And so has its analogue in the collapse of the surpernova core. As the core spins up in the collapse, it too should be prone to vortex formation, and the neutrinos will

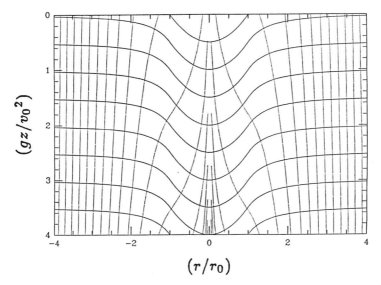

FIGURE 5. Focusing effect of a low-pressure vortex. The *solid lines* are contours of $C_p T/v_0^2$, as given by (**28**). The contour interval is 0.5. The *dotted lines* are flow lines for ∇T, which by (**27**) are approximately flow lines for the radiative energy flux F. The contour interval is arbitrary.

beam out. In all these questions, both the Jovian and the solar observations will guide our thinking. The effects of rapid rotation are evident in the former and the role of magnetic fields are crucial in the latter.

6. CONCLUSION

We invest great scientific effort to explain what we see in nature. But sometimes, when we have understood better, we realize that some of these things are generic and that if we did see them, *then* we would then have some real explaining to do. Don Giovanni should not have been so surprised, as the flames of hell reached up for him, to observe that they were full of vorticity. As for ourselves, we have now learned that whenever there is an unstable zonal wind profile we must expect the generation of strong vortices. That is the lesson of the Jovian planets. We are trying to apply it to hot stars. But there is also baroclinic vortex generation, as we know from terrestrial weather. This latter process is more difficult to think about, especially against a background of strong thermal turbulence.

A similar dichotomy is expected in stars. The solar-type stars—at least the ones that are not rapid rotators—are more like the terrestrial planets in our parallelism. They have vortices, it now appears, but those probably arise from the complexity of convection, and the ones seen thus far have had negligible help from the global rotation. On the other hand, in studying the magnetic flux tubes that cause the sunspots, we may have overlooked a good bet in not using the planetary observations as a guide.

In the planetary case, the Jovian vortices are caused by zonal shear flows. The main theoretical problem is to predict the nature of zonal winds on Jovian planets. What are the sources of the shear? Or more technically, what are the sources of potential vorticity extrema, and how is the apparent tendency for potential vorticity to homogenize overcome? It seems generally agreed that if we could come up with a way to maintain unstable zonal winds, we would produce an atmosphere covered with large, long-lived vortices. As to the sun, it has been suggested that the spots arise in a strong toroidal field *below* the convection zone. Undoubtedly, we need to understand more about the spatial structure of this field, especially of the distribution of the hydromagnetic analogue of potential vorticity.

In the case of hot stars, which correspond to the Jovian planets in the planetary metaphor, we accept the dogma that rapid rotation is associated with strong zonal winds. This is the most uncertain part of the story, for we have no real theory of such winds as yet. But once they are accepted, the shear flow instabilities may be assumed to produce vortices. What we have done here is to accept all that in order to anticipate how a rapidly rotating star covered with vortices may look if it is also quite luminous. The prospects of a hot stellar atmosphere crisscrossed with strong beams of radiation like the opening of a shopping mall make for some complicated photogasdynamics. This is a situation that ought to seed the powerful winds seen around such stars. It should certainly be rewarding to pursue this problem in the study of hot accretion disks and the analogous dynamics of a neutrino–gas mixture.

ACKNOWLEDGMENTS

In developing the ideas presented here we have benefitted from very helpful interactions with Joseph Casinelli, Peter Gierasch, Andrew Ingersoll, E. T. Scharlemann, and Ron Tam.

REFERENCES

1. INGERSOLL, A. P., R. F. BEEBE, B. J. CONRATH & G. E. HUNT. 1984. Structure and dynamics of Saturn's atmosphere. *In* Saturn, T. Gehrels and M. S. Matthews, Eds.: 195–238. University of Arizona Press. Tucson.
2. WILLIAMS, G. P. 1985. Jovian and comparative atmospheric modeling. Adv. Geophys. **28A:** 381–429.

3. FLASAR, F. M. 1986. Global dynamics and thermal structure of Jupiter's atmosphere. Icarus 65: 280–303.
4. EMANUEL, K. 1988. Influence of convection on large scale circulations. Summer Study Program in Geophysical Fluid Dynamics, Tech. Rep. WHOI-89-26. Woods Hole Oceanographic Institution, Woods Hole, Mass.
5. BRANDT, P. N., G. B. SCHARMER, S. H. FERGUSON, R. A. SHINE, T. D. TARBELL & A. M. TITLE. 1988. Nature 335: 238.
6. GRAHAM, E. 1977. In Problems of Stellar Convection, E. A. Spiegel and J.-P. Zahn, Eds.: 151. Springer-Verlag. Berlin/New York.
7. CASINELLI, J. P. 1985. Evidence for non-radiative activity in hot stars. In The Origin of Nonradiative Heating/Momentum in Hot Stars, A. B. Underhill and A. G. Michalitsianos, Eds.: 2–23. NASA 2358.
8. BJERKNES, V. 1926. Astrophys. J. 64: 93.
9. MacLow, M.-M. & A. P. INGERSOLL. 1986. Merging of vortices in the atmosphere of Jupiter: An analysis of Voyager images. Icarus 65: 353–369.
10. RHINES, P. B. 1975. Waves and turbulence on a beta-plane. J. Fluid Mech. 69: 417–443.
11. ———. 1979. Geostrophic turbulence. Annu. Rev. Fluid Mech. 11: 401–441.
12. INGERSOLL, A. P. & P. G. CUONG. 1981. Numerical model of long-lived Jovian vortices. J. Atmos. Sci. 38: 2067–2076.
13. McWILLIAMS, J. C. 1984. The emergence of isolated coherent vortices in turbulent flow. J. Fluid Mech. 146: 21–43.
14. SOMMERIA, J., S. D. MEYERS & H. L. SWINNEY. 1988. Laboratory simulation of Jupiter's Great Red Spot. Nature 331: 689–693.
15. MARCUS, P. S. 1988. Numerical simulation of Jupiter's Great Red Spot. Nature 331: 693–696.
16. NEZLIN, M. V. & G. G. SUTYRIN. 1989. Long-lived solitary anticyclones in the planetary atmospheres and oceans, in laboratory experiments and in theory. In Mesoscale—Synoptic Coherent Structures in Geophysical Turbulence. Elsevier Oceanography Series, Vol. 50: 701–719. Elsevier. Amsterdam/New York.
17. WILLIAMS, G. P. & R. J. WILSON. 1988. The stability and genesis of Rossby vortices. J. Atmos. Sci. 45: 207–241.
18. DOWLING, T. E. & A. P. INGERSOLL. 1989. Jupiter's Great Red Spot as a shallow water system. J. Atmos. Sci. 46: 3256–3278.
19. SMITH, B. A., et al. 1989. Voyager 2 at Neptune: Imaging science results. Science 246: 1422–1449.
20. INGERSOLL, A. P. 1988. Models of Jovian vortices. Nature 331: 654–655.
21. GILL, A. E. 1982. Atmosphere-ocean Dynamics. Academic Press. New York.
22. PEDLOSKY, J. 1987. Geophysical Fluid Dynamics, 2d ed. Springer-Verlag. New York/Berlin.
23. ERTEL, H. 1942. Ein neuer hydrodynamischer Wirbesatz. Meterolol. Z. 59: 277–281.
24. FLIERL, G. R. 1987. Isolated eddy models in geophysics. Annu. Rev. Fluid Mech. 19: 493–530.
25. DOWLING, T. E. & A. P. INGERSOLL. 1988. Potential vorticity and layer thickness variations in the flow around Jupiter's Great Red Spot and White Oval BC. J. Atmos. Sci. 45: 1380–1396.
26. GENT, P. R. & J. C. McWILLIAMS. 1983. Regimes of validity for balanced models. Dyn. Atmos. Oceans 7: 167–183.
27. SPIEGEL, E. A. 1972. A History of Solar Rotation, Physics of the Solar System, S. I. Rasool, Ed.: 61. Scientific and Technical Information Office Tech. Rep., NASA SP-300, NASA, Washington, D.C.
28. ANTIPOV, S. V., M. V. NEZLIN, E. N. SNEZHKIN & A. S. TRUBNIKOV. 1986. Rossby autosoliton and stationary model of the jovian Great Red Spot. Nature 323: 238–240.
29. RHINES, P. B. & W. R. YOUNG. 1982. Homogenization of potential vorticity in planetary gyres. J. Fluid Mech. 122: 347–367.
30. ACHTERBERG, R. K. & A. P. INGERSOLL. 1989. A normal mode approach to Jovian atmospheric dynamics. J. Atmos. Sci. 46: 2448–2462.
31. FLIERL, G. R. 1984. Rossby wave radiation from a strongly nonlinear warm eddy. J. Phys. Ocean. 14: 47–58.

32. READ, P. L. & R. HIDE. 1983. Long-lived eddies in the laboratory and in the atmospheres of Jupiter and Saturn. Nature **302:** 126–129.
33. ———. 1984. An isolated baroclinic eddy as a laboratory analogue of the Great Red Spot on Jupiter? Nature **308:** 45–49.
34. READ, P. L. 1986. Stable, baroclinic eddies on Jupiter and Saturn: A laboratory analogue and some observational tests. Icarus **65:** 304–334.
35. BROWN, T. M., J. CHRISTENSEN-DALSGAARD, W. DZIEMBOWSKI, P. GOODE & D. O. GOUGH. 1989. Inferring the sun's internal velocity from observed p-mode splittings. Astrophys. J. **343:** 526–546.
36. SMITH, B. A., *et al.* 1979. The Galilean satellites and Jupiter: Voyager 2 imaging science results. Science **206:** 927–950.
37. SROMOVSKY, L. A., H. E. REVERCOMB, R. J. KRAUSS & V. E. SUOMI. 1983. Voyager 2 observations of Saturn's northern mid-latitude cloud features: Morphology, motions and evolution. J. Geophys. Res. **88:** 8650–8666.
38. DOWLING, T. E. & P. J. GIERASCH. 1989. Cyclones and moist convection on Jovian planets. Bull. Am. Astron. Soc. **21:** 946.
39. BEEBE, R. F., G. S. ORTON & R. A. WEST. 1989. Time-variable nature of the Jovian cloud properties and thermal structure: An observational perspective. *In* Proceedings of the Workshop on Time-variable Phenomena in the Jovian System, M. J. S. Belton, R. A. West, and J. Rahe, Eds. NASA SP-494. NASA, Washington, D.C.
40. PEEK, B. M. 1981. The Planet Jupiter, rev. ed. Faber & Faber. London.
41. WILLIAMS, G. P. & T. YAMAGATA. 1984. Geostrophic regimes, intermediate solitary vortices and Jovian eddies. J. Atmos. Sci. **41:** 453–478.
42. SIMON, G. W. & N. O. WEISS. 1989. Astrophys. J. **345:** 1060–1078.
43. DEUBNER, F.-L. & D. O. GOUGH. 1984. Annu. Rev. Astron. Astrophys. **22:** 593.
44. COWLING, T. G. 1953. *In* The Sun, chap. 8, G. P. Kuiper, Ed., University of Chicago Press. Chicago.
45. AKOSOKA, S.-I. 1985. Vortical distribution of sunspots. Planet. Space Sci. **33:** 275–277.
46. LEDOUX, P., M. SCHWARZSCHILD & E. A. SPIEGEL. 1961. On the spectrum of turbulent convection. Astrophys. J. **133:** 184.
47. McEWAN, A. D. 1976. Nature **260:** 126–128.
48. HOPFINGER, E. J. & F. K. BROWAND. 1982. Vortex solitary waves in a rotating, turbulent flow. Nature **295:** 393–395.
49. SCHWARZSCHILD, M. 1961. Convection in stars. Astrophys. J. **134:** 1.
50. KRAICHNAN, R. H. 1961. Private communication.
51. SPIEGEL, E. A. 1987. Chaos: A mixed metaphor for turbulence. Proc. R. Soc. London, Ser. A **413:** 87.
52. BUSSE, F. H. 1981. Transition to turbulence in Rayleigh-Bénard convection. *In* Hydrodynamic Instabilities and the Transition to Turbulence. H. L. Swinney and J. P. Gollub, Eds.: 97–133. Springer-Verlag. New York/Berlin.
53. CHANDRASEKHAR, S. 1961. Hydrodynamic and Hydromagnetic Stability. Clarendon Press. Oxford. (Reprinted by Dover. New York. 1968.)
54. ZIPPELIUS, A. & E. D. SIGGIA. 1983. Phys. Fluids **26:** 2905.
55. MASSAGUER, J. M. & I. MERCADER. 1988. Instability of swirl in low-Prandtl-number thermal convection. J. Fluid Mech. **189:** 367–395.
56. STIX, M. 1989. The Sun. Springer-Verlag. New York/Berlin.
57. ZELIK, M., D. S. HALL, P. A. FELDMAN & F. WALTER. 1979. The strange RS Canum Venaticorum binary stars. Sky Telesc. **57:** 132–140.
58. STRUVE O. & S.-S. HUANG. 1960. *In* Stellar Atmospheres, J. L. Greenstein, Ed.: 300. University of Chicago Press. Chicago.
59. Science **244:** 31.
60. LIMAYE, S. S. 1986. Jupiter: New estimates of the mean zonal flow at the cloud level. Icarus **65:** 335–352.
61. SMITH, B. A., *et al.* 1986. Voyager 2 in the Uranian system: Imaging science results. Science **233:** 43–64.
62. HAMMEL, H. B., *et al.* 1989. Neptune's wind speeds obtained by tracking clouds in Voyager images. Science **245:** 1367–1369.

Fluid Dynamics of Astrophysical Jets[a]

MICHAEL L. NORMAN

National Center for Supercomputing Applications
and
Department of Astronomy
University of Illinois at Urbana-Champaign
Urbana, Illinois 61801

1. INTRODUCTION

The study of astrophysical jets, both theoretically and observationally, has moved into the mainstream of astronomical research in recent years because of their apparent ubiquity in the universe, their mysterious nature and origin, and their intrinsic beauty as revealed by powerful radio interferometers, sensitive digital photometers, and supercomputer simulations. The famous optical jet in the giant elliptical galaxy M87 discovered by Curtis in 1918 remained an astronomical curiosity until the early 1970s, whereupon extragalactic radio jets began to be detected in a large number of 3C radio galaxies. A decade and a half of intensive observational work by many investigators has revealed over 200 radio jets that are associated with active galaxies of all types, including Seyfert galaxies, elliptical galaxies, and quasars.[1] Many radio jets have been mapped in great detail[2-4] with the National Radio Astronomy Observatory's (NRAO) Very Large Array (VLA) radio telescope in Socorro, New Mexico, and their essential structural properties have been deduced.[1]

Early in the 1980s, the discovery of high-velocity, ionized jets in star-forming regions in our own galaxy,[5,6] as well as in the compact binary system SS 433[7] broadened our perspective once again by showing that jets could be produced in systems with characteristic mass and length scales one million times smaller than active galactic nuclei, yet the jets display similar structural properties such as degree of collimation, length-to-width ratios, and morphological substructure. Subsequent optical surveys of star-forming regions have turned up some 20 examples with new ones continuing to be discovered.[8] Thus, jets have emerged as a new class of astronomical *object* worthy of study in their own right.

The significance of jets as part of a *process* is only now emerging through spectroscopic studies of star-forming regions in the optical, infrared, and millimeter wavelengths. A number of lines of evidence, both observational and theoretical, suggest that protostellar jets and associated bipolar molecular outflows may be a necessary by-product of the star-formation process.[9] The current view is that protostellar disk accretion and jet ejection are physically linked, with the jets and molecular outflows carrying off the excess angular momentum that must be lost by accreting material if it is to be incorporated into the protostar. In other words, jets

[a] A number of the calculations reported here were performed at the National Center for Supercomputing Applications, University of Illinois at Urbana-Champaign, under partial support from National Science Foundation Grants AST-8611511 and AST-8516921, and by EPSCoR Grant RII-8610669.

217

may indicate nature's solution of the well-known angular-momentum problem in star formation. It is likely that jets from AGN derive from the same or similar accretion processes. If this is so, then the explanation for why jets exist on such vastly different scales is that the physics of disk accretion is similar from protostars to supermassive black holes.[10]

In this review, I approach jets as *physical systems* distinct from their method of generation. This approach reflects both an historical and an observational bias. Historically, the acceleration and collimation of jets has always been treated as a separate problem from the structure and evolution of jets. Observationally, the problems separate as one can spatially resolve both protostellar and extragalactic jets, whereas their generation region remains unresolved. This division may turn out to be physically untenable; jet properties may be inextricably linked to the properties of the "central engine." This possibility, however, has been used to justify such a division insofar as jet studies may provide useful constraints on the acceleration mechanism. An alternate justification, and one that motivates much of the current work on extragalactic jet simulations summarized here, is that jet morphology may be a useful probe of the galactic environment. In fact, I would make a stronger statement: by modeling situations where radio jets and their environments strongly interact, such as in head–tail radio galaxies or disrupting jets, one can simultaneously constrain both jet and ambient media that are only poorly constrained by independent observations.

Consistent with the theme of this conference, I will focus on various fluid dynamical aspects of astrophysical jets rather than their observational characteristics. Good observational reviews of both the protostellar and extragalactic jet varieties can be found in Mundt[11] and Bridle and Perley,[1] respectively. Furthermore, I will not attempt to review the many theoretical aspects of extragalactic radio source physics that are outside the realm of fluid dynamics (e.g., particle acceleration), which are nonetheless essential to understanding the phenomenon. For this I refer the reader to the monograph by Begelman, Rees, and Blandford.[12] Rather, I will focus on the nonlinear fluid dynamical structures, instabilities, and processes revealed by numerical simulations that may be relevant to the interpretation of both extragalactic radio jets and ionized protostellar jets. Numerical simulations are required because of the inherent nonlinearities of the equations of fluid dynamics, and the fact that parameter regimes of astrophysical interest are generally experimentally inaccessible. Wherever possible, the results of simulations are related to simple analytic theory or perturbation theory.

In Section 2 I review the properties of the "Standard Model"—that of an adiabatic, light, supersonic gas jet penetrating a constant density and pressure background—as a means of introducing the basic flow structures present as well as providing a point of reference for jet simulations where inlet, ambient, thermodynamic, and dynamical conditions are varied. Most numerical simulations to date have been performed in two spatial dimensions due to limitations in computer memory and speed. Nevertheless, by exploiting the complementary coordinate geometries of cylindrical (axisymmetry) and Cartesian (slab symmetry), one can explore a wide variety of modes of instability that would afflict a three-dimensional jet. In Section 3 I discuss the topic of jet stability and summarize the results of such simulations as well as their agreement with linear perturbation theory. The effects of gradients and jumps in the ambient medium on jet structure and stability are discussed in Section 4. In Section 5 I summarize the results of simulations where magnetic stresses and radiative cooling have been added to the equations of motion, as motivated by a consideration of quasar jets and protostellar jets, respectively. Finally, in Section 6, I review what we have learned from several attempts to model jets in three dimensions, and I pose some questions deserving of further study.

2. BASIC FLOW STRUCTURES: THE STANDARD MODEL

2.1. Definition of the Standard Model

The Standard Model consists of the time-dependent solution to the following initial-boundary value problem in ideal gas dynamics (an informative discussion of the applicability of ideal gas dynamics to astrophysical jets can be found in Begelman, Rees, and Blandford[12]). Consider a rectangular domain filled with a uniform, stationary gas characteristic of the ambient medium through which the jet propagates. The jet is continuously injected through an inlet on the boundary of this region with a fixed density, pressure, and velocity. All other points on the boundary are treated as either reflecting (i.e., zero normal velocity) or transmitting (i.e., zero normal gradient) boundaries, depending upon the assumed symmetry of the problem and the boundary in question. Axisymmetry is assumed in the Standard Model, thus, the boundary coinciding with the jet axis is taken to be a reflecting boundary. All other boundary segments are taken to be transmitting. Both gases are assumed to obey the ideal gas equation of state, with adiabatic exponent 5/3. Solutions to this problem are obtained by direct numerical integration of the equations of ideal gas dynamics, for which there are now reliable methods[13,14] and adequate computer power for the two-dimensional problem. The solutions can be parameterized by the ratio of jet density to ambient density, designated η, the ratio of the jet pressure to the ambient pressure, designated K, and the jet Mach number in units of the internal sound speed, designated M. A common choice of parameters that I use to denote the Standard Model is ($\eta = 0.1, M = 6, K = 1$). Numerical simulations of the Standard Model have been performed by a long list of authors[15-26] with essentially identical results. The main structural features elucidated by these simulations are shown schematically in FIGURE 1. A computer animation of the Standard Model is included in Segment 1 of the videotape (see Nonlinear Astrophysical Fluid Dynamics: The Video, in this volume, for availability and ordering information).

2.2. The Mean Flow

Loosely speaking, the flow can be divided into three regions: (1) the incident supersonic jet, (2) the jet-termination shock wave (the "working surface"), and (3) the cocoon. A given fluid element passes through these three regions sequentially. This basic flow structure was inferred on purely theoretical grounds by Blandford and Rees,[27] and can be understood on the basis of mass, momentum, and energy conservation. By balancing ram pressures at the jet head, one can derive the well-known result that $V_{\text{head}} = \eta^{1/2}V_{\text{jet}}/(1 + \eta^{1/2})$, where V_{head} and V_{jet} are the head and jet velocities, respectively. For a low-density jet ($\eta \ll 1$), this reduces to $V_{head} = \eta^{1/2}V_{jet} \ll V_{jet}$, that is, the jet material is constantly catching up to the jet head. In this case, the relative velocity $V_{jet} - V_{head}$ can be quite large; in fact, it is generally supersonic with respect to the jet sound speed. Consequently, the jet flow is decelerated at a shock wave at the end of the jet. At the working surface, the jet kinetic energy is converted into internal energy, producing a high-pressure region whose pressure is equal to the incident ram pressure of the jet $\rho_{jet}(V_{jet} - V_{head})$.[2] The jet head advances supersonically into the ambient medium; hence, it drives a bow shock ahead of itself. This static pressure is balanced in the forward direction by the ram pressure of the oncoming ambient medium $\rho_{ext}V_{head}^2$, where ρ_{jet} and ρ_{ext} are the densities of the jet and external medium, respectively. The pressure difference between the working surface and the ambient medium drives a vigorous flow

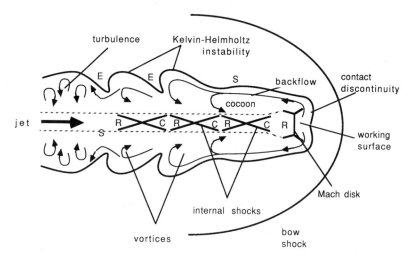

FIGURE 1. Schematic flow structures of a light ($\eta < 1$), supersonic jet as deduced from two-dimensional axisymmetric, time-dependent, numerical simulations. Legend for flow regions as follows: R = rarefaction; C = compression; S = shear; E = entrainment.

sideways and back, inflating a cocoon with shocked jet gas. From mass conservation, the cocoon is inflated at a rate given by $\rho_{jet}(V_{jet} - V_{head})A_{jet}$, where A_{jet} is the jet area. From energy conservation and assuming a steady flow (see discussion below), one can derive the gross properties of the cocoon.[28] From the preceding considerations, it is easy to see that light jets ($\eta \ll 1$) should have substantial cocoons, and that heavy jets ($\eta \gg 1$) should have virtually no cocoon (naked jets).

Observationally, one associates the high-pressure region behind the terminal shock with the hot spots in extragalactic radio jets, and the cocoon with the radio lobe. That the observed lobe diameters greatly exceed the jet diameters implies $\eta \ll 1$; that is, radio jets are extremely diffuse compared to the mean density of the intergalactic medium. The situation with protostellar jets, while analogous, differs in some important respects, as will be discussed in Section 5.2.

2.3. Nonlinear Structures

The early simulations of Norman et al.[15-17,19] confirmed the basic flow structure and scaling arguments set forth by Blandford and Rees,[27] but in addition, revealed a wealth of spatial and temporal structure *superimposed* on the general flow that had not been predicted. Subsequent investigations[18,20-26] have refined and qualified the representation given in FIGURE 1. Following the fluid path, it is found that the incident supersonic jet contains internal shock waves that are oblique. Owing to the assumed axisymmetry, they are geometrically biconical. These shocks are generated in response to perturbations applied to the jet from fluid motions in the cocoon, or through the growth and saturation of Kelvin–Helmholtz pinch instabilities. The effect of these shocks is to create a series of high- and low-pressure zones along the jet, which may be the origin of emission knots in astrophysical jets[15,29] (see Section 3.2).

As can be seen on the videotape, the terminal shock is unsteady, oscillating between a characteristic Mach disk configuration and an X-shaped configuration. The origin of this unsteadiness is not completely understood, but seems to be related to the quasi-periodic creation and "shedding" of large-scale vortices at the working surface.[15,20] These vortices are transported by the mean flow away from the working surface and into the cocoon, giving it a complex, turbulent structure. As the vortices move, they modulate the incident jet and terminal shock structure. Vorticity is created at the terminal shock in abundance due to its nonuniform structure; for example, at the triple point where the incident, reflected, and Mach disk shocks meet. Thus, it is not surprising that this vorticity is shed in discrete events, as it is in high Reynolds-number flow about a bluff body to produce the well-known Von Karman vortex street. In principle, the Reynolds number of these simulated jets is infinity as one is solving the inviscid equations. In practice, numerical dissipation arising from truncation errors in the difference equations solved introduces a grid viscosity that is nonzero. The grid viscosity is determined by the order of accuracy of the numerical algorithm and the number of grid points used to resolve the character-istic length entering into the definition of the Reynolds number. The effective Reynolds number is thus increased as more zones and more accurate methods are applied. Attempts to estimate the Reynolds number in simulations of compressible convection using the very accurate piecewise parabolic method (PPM) (see the paper by Porter et al. in this volume) yields a scaling law which, when applied to the present problem and methods, yields Reynolds numbers in the range 10^3–10^4—sufficiently large to exclude steady, laminar flow. Kössl and Müller[26] performed a resolution study of the Standard Model with as many as 200 zones across the jet radius, and showed that terminal shock oscillations were enhanced with increasing resolution. High-resolution simulations using PPM also show unsteadiness at the head, and reveal a greater variety of vortex shedding events reminiscent of the turbulent transition regime.

2.4. Surface Instabilities

Since the boundary of the jet is a free shear surface, it is subject to the Kelvin–Helmholtz instability. The working surface oscillations provide an order unity distortion to the contact discontinuity near the apex of the jet; these perturbations are amplified by the Kelvin–Helmholtz instability as they are convected back along the jet. Numerical simulations show that the growth of these instabilities is more pronounced at lower Mach numbers, as one would expect, and leads to considerable entrainment of ambient material into the cocoon. This effect has yet to be quantified; indeed, it may be meaningless to do so within the context of two-dimensional simulations, as many more unstable wave modes can be excited in three dimensions, which may contribute to the entrainment rate. This topic will be revisited in Section 6.

3. STABILITY OF PROPAGATION

Observations of extragalactic radio jets provide many examples where well-collimated jets propagate stably for many tens to one hundred times their diameter without bending or flaring (e.g., NGC6251).[2] In other cases, radio jets are seen to develop sinusoidal wiggles beyond a straight section (e.g., 3C 449).[30] In yet other cases, straight jets begin to meander as they enter the radio lobe and approach the radio hot spot (e.g., Cyg A).[3] Jets in head–tail radio galaxies miraculously bend through 90° before disrupting into turbulent looking radio tails (e.g., NGC 1265).[31] Finally, jets in high-luminosity radio galaxies and quasars, as well as in some protostellar jets, display a train of emission knots that are quite regularly spaced (e.g., 0800 + 608; HH34 jet).[32,33] These and other observed morphologies have stimulated a large number of theoretical studies of jet stability addressing the following basic questions: (1) What determines how far a jet can propagate? (2) Which modes of instability are present and how do they manifest? (3) Can observed morphologies be used to constrain the jets' physical parameters (density, velocity, etc.)? The last question is of particular importance as applied to extragalactic jets, where, due to the continuum nature of synchrotron emission, one has no direct measure of these quantities. In this section, I will first summarize the basic findings of numerous linear analyses of jet stability by many authors. These studies provide a framework for the subsequent discussion of a series on nonlinear studies carried out by collaborators and myself that provide some answers to the questions above.

3.1. Summary of Linear Stability Analyses

Linear stability analyses of the Kelvin–Helmholtz instability for the case of a cylindrical supersonic beam in a constant atmosphere have been performed by many authors.[34-36] Their basic findings are that: (1) there exist modes of instability with significant spatial growth rates (i.e., e-folding lengths) at all Mach numbers; (2) these modes can be classified as to whether they are axisymmetric (i.e., pinching modes) or nonaxisymmetric (i.e., helical modes), and (3) both symmetries exhibit a family of unstable modes characterized by the number of nodes in the radial eigenfunction used in the perturbation analysis. The lowest order mode of each symmetry is referred to as the fundamental mode, while the overtones are referred to reflection modes. Reflection modes derive their name from the special property of a supersonic shear surface, which is that for certain angles of incidence, a plane linear wave will be amplified upon reflection from the interface. A reflection-mode instability occurs in a supersonic beam when waves repeatedly reflect between the walls of the jet, growing in amplitude with each bounce. A thorough discussion of this physics in the context of radio jets is given in Payne and Cohn.[36]

FIGURE 2(a) shows a typical solution of the complex dispersion relation resulting from a spatial mode analysis (complex wavenumber versus real frequency) for an $\eta = 0.1$, $M = 3$ jet (from Norman and Hardee[37]). Here the fundamental and first reflection mode is shown for both the pinch and helical instability. Notice that all but the fundamental pinch mode have comparable spatial growth rates over a wide range of driving frequencies. The higher order reflection modes generally peak at higher frequencies with slightly higher growth rates. This implies that all modes with the exception of the fundamental pinch mode may be present at once in a real jet. Therefore, we must turn to numerical simulation to see how they manifest in the nonlinear regime and which modes dominate the jet's structure and evolution.

3.2. Nonlinear Pinch Instabilities

Axisymmetric simulations of naked jets obeying the supersonic criterion $V_{jet} > C_{jet} + C_{ext}$ show that pinch modes saturate through the formation of weak internal shock waves.[17] These shocks are oblique and reflect on the axis of the jet describing a geometrically biconical structure. Generally, these biconical shocks exhibit a cellular structure similar in appearance to the "shock diamonds" seen in laboratory under expanded jets.[38] The shock cells have a characteristic wavelength of roughly five jet radii, although this result is weakly Mach number dependent. Unlike the laboratory situation, where the shock diamonds are attached to the orifice, the simulated shock diamonds move downstream with speeds approximately $V_{jet}-C_{jet}$; that is, the velocity of the jet relative to the shock pattern is roughly sonic. A detailed analysis by Payne and Cohn[36] indicates that these shocks are likely the nonlinear saturated state of the first reflection-mode pinch instability. As mentioned earlier, these shocks create regions of high pressure just downstream of their point of reflection, which may account for the bright emission knots seen in astrophysical jets.[29]

3.3. Nonlinear Kink Instabilities

A more fundamental concern than the origin of knots is whether the growth of nonaxisymmetric modes of the Kelvin–Helmholtz instability, such as the fundamental helical mode, can destroy the directional stability of astrophysical jets. At the present time, it is not feasible to perform three-dimensional simulations of sufficient accuracy to make a comparison with perturbation theory meaningful. Instead, Phil Hardee and I undertook a series of numerical experiments[37,39,40] on two-dimensional slab jets that we believe capture the essential physics. A slab jet is spatially resolved along two Cartesian axes and is infinite in extent in the third direction. This symmetry permits the jet to develop bends and kinks that are strictly prohibited in axisymmetric

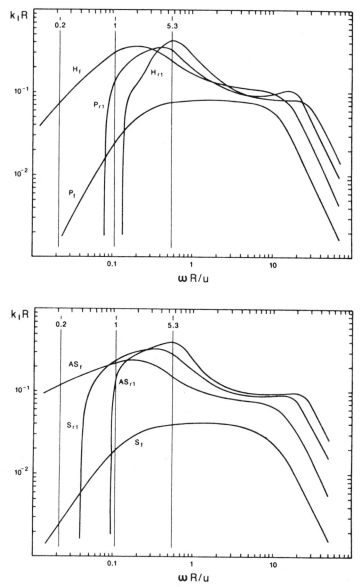

FIGURE 2. Comparison of the linear response of a three-dimensional cylindrical jet with a two-dimensional slab jet with $\eta = 0.1$, $M = 3$. (From Hardee and Norman.[39] Reproduced by permission.) (**a**) Spatial growth rate, $k_i R$, as a function of the frequency $\omega R/u$, of the pinch fundamental, P_f, and first reflection, P_{r1}, modes, and of the helical fundamental, H_f, and first reflection, H_{r1}, modes for the cylindrical jet. (**b**) Spatial growth rate, $k_i R$, as a function of the frequency $\omega R/u$, of the symmetric fundamental, S_f, and first reflection, S_{r1}, modes, and of the asymmetric fundamental, AS_f, and first reflection, AS_{r1}, modes for the slab jet.

calculations. The physical justification for this approach is contained in FIGURE 2(b), which shows the solution of the dispersion relation for linear perturbations on a slab jet of identical parameters to that in FIGURE 2(a). The modes that correspond to the pinch (P) and helical (H) modes of FIGURE 2(a) are labeled symmetric (S) and antisymmetric (AS), respectively, for both the fundamental (f) and first reflection modes (r1). The detailed similarity in both the modal structure and spatial growth rates between the cylindrical and the slab jets, first pointed out by Ferrari et al.,[41] validates our approach.

The setup of the problem and a computer animation of the results can be found in Segment 2 of the videotape that complements this paper. Briefly, a slab jet is initialized spanning the computational domain, flowing from left to right and in pressure balance with the uniform, ambient gas. A periodic, transverse velocity perturbation is applied to the gas entering at left with the frequency as a free parameter. Typically, several frequencies around and including the resonant frequency as determined by perturbation theory were tried. The video animation reveals three phases of evolution of the slab: (1) spatial growth of a transient kink disturbance to perceptible amplitude as it is convected downstream; (2) subsequent disruption of the jet channel due to the nonlinear growth of the kink; (3) the inflation of a gaseous lobe and the emergence of a well-defined bow shock as the newly created jet head advances through the ambient gas. The lobe is supplied by the jet as in the Standard Model; however, the jet becomes a driven oscillator and flaps through several jet widths as it approaches the working surface.

By measuring the wavelengths and pattern speeds of the sinusoids that define the jet channel at any instant for different jet parameters and driving frequencies, we can directly compare with perturbation theory. We find good agreement at late times by interpreting the flapping as the driven response of the jet to the fundamental kink mode.[37] Generally, we find that fundamental kink instability reaches large amplitude in 6–7 resonant e-folding lengths as determined by perturbation theory; for example, about 10 MR_{jet}. The early disruptive kink was found to agree well with the first antisymmetric reflection mode in terms of its wavelength, but not so well in terms of its pattern speed. A subsequent analysis[42] has revealed a new resonance mode that can be excited by an arbitrary disturbance, whose characteristics agree well with the numerical results.

3.4. Jet Deflection

Although the simulations previously described are not realistic models of radio source formation due to their starting conditions, one would expect on general grounds to see similar phenomena (jet flapping) in a propagating jet simulation. If this occurs in a real jet, then we may have an answer to why some jets meander as they approach the hot spot. To investigate this, Phil Hardee and I performed slab jet simulations[43] where the jet is propagated into the ambient gas rather than established as a preexisting channel. As before, a velocity perturbation is applied at the resonant frequency of the fundamental mode to break lateral symmetry. The results of these simulations are shown on videotape Segment 3 and are summarized in FIGURE 3.

The slab jet passes through three distinct phases as it evolves that we have called the symmetric, the asymmetric, and the choatic phases. During the symmetric phase, the flow structures are symmetric about the midplane and resemble those of the Standard Model in all respects, including the formation of a Mach disk terminal shock, internal jet shocks, and vortices in the cocoon. The symmetric phase lasts until $\tau V_{jet}/C_{ext} \approx 17$, where τ is the age of the jet, whereupon asymmetries appear. These asymmetries consist of detectable bends in the jet, oblique internal shock waves whiten the jet and at its termination, and cocoon vortices that are more pronounced on one side of the jet than the other. During this phase, wiggles in the jet bear some relationship to the input perturbations. Beyond $\tau V_{jet}/C_{ext} \approx 25$, the chaotic phase sets in, where this is no longer true. The chaotic phase is characterized by a strong buffeting of the jet due to large vortices in the cocoon, which are not symmetrically

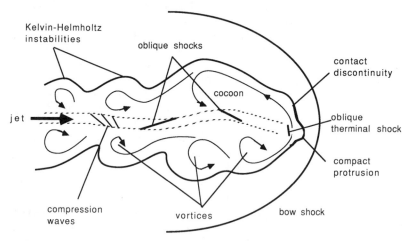

FIGURE 3. Schematic flow structures of a light ($\eta < 1$), supersonic jet as deduced from two-dimensional slab-symmetric, time-dependent, numerical simulations.

disposed about the jet. In other words, the cocoon-driven biconical shocks seen in the Standard Model have been replaced by oblique shocks that deflect the jet. As can be seen in the video animations, jet deflection sometimes occurs well upstream of the tip of the cocoon, which results in a highly nonstationary working surface. In some instances, the jet's point on impingement paints across the inner edge of the cocoon; in other instances, the jet is chopped off well behind the leading edge and then propagates to a new point of impact on the cocoon surface.

These simulations reveal a physical mechanism that may underlie the "dentist drill" model advanced by Scheuer[44] to explain multiple hot spots in certain powerful radio doubles. One uncertainty that remains to be explored is whether the assumed slab symmetry creates or amplifies jet buffeting beyond what would be expected in a three-dimensional jet. Three-dimensional numerical simulations that should be able to determine this are becoming feasible now.

4. JETS IN NONUNIFORM ATMOSPHERES

4.1. Effect of Atmospheric Gradients

As jets propagate outward from the centers of galaxies they expand as they encounter a declining ambient pressure. In two well-studied extragalactic radio jets—NGC 6251 (reference 2) and 3C 120 (reference 45)—this expansion can be tracked over six orders of magnitude in length scale and the atmospheric pressure gradient can be inferred. The ambient pressure is found to obey a power law in radius with an exponent of approximately two.[45] The hydrodynamically interesting question is whether jet expansion leads to enhanced stability. Linear analyses[35,46–49] have shown that jet expansion and jet cooling slow the growth of perturbations and partially stabilize the jet. In particular, the wavelength and spatial growth length of helical twisting should scale with Mach number and jet radius.

To see whether these linear predictions carry over into the nonlinear regime, Phil Hardee and I ran a series of slab jet numerical experiments identical to those described in Section 3.3; however, with a variety of atmospheric gradients including isothermal pressure power law, isobaric temperature power law, and pressure power law with a temperature jump. In each case, the slab jet was initialized in pressure balance with its environment, and hence possessed a nonzero opening angle. We find that jets are stabilized by jet expansions as predicted by the linear analysis. We also find that an expanding jet can be destabilized by a positive temperature gradient of temperature jump in the surrounding medium that lowers the Mach number defined by the external sound speed. This work is described in Hardee et al.[50]

4.2. Atmospheric Discontinuities and Jet Disruption

Observations show that the well-collimated jets in moderate luminosity ($< 10^{25}$ W Hz^{-1} at 20-cm wavelength) radio galaxies can flare dramatically in a few jet diameters and with opening angles up to 90° into diffuse lobes or tails.[51,52] Quite often, this flaring is seen to occur in both radio jets as they enter the galactic halo suggesting an environmental origin. By analogy with structures seen in the laboratory, one could interpret the collimated jets as moderately supersonic flows, and the lobes or tails as subsonic plumes that are subject to turbulent broadening and entrainment of the ambient medium. Sudden supersonic/subsonic transitions have not been reproduced in numerical simulations of jets in atmospheres where the pressure is constant or smoothly varying. Thus, one is motivated to consider the effects of an ambient pressure jump on jet propagation. Such a jump could form in galactic atmospheres through two possible mechanisms: (1) at a shock wave created by a galactic wind encountering the intergalactic medium; or (2) at the "pressure wall" of a transonic cooling inflow.[53]

My collaborators and I have investigated both scenarios and have found that jet flaring does occur under rather broad and understandable circumstances. Norman, Burns, and Sulkanen[54] simulated the passage of a slab jet across an inward-facing shock wave (i.e., pressure jump is positive, as it would be crossing a galactic wind shock), and found that jet disruption and flaring occur provided that the Mach number of the jet was less than the Mach number of the wind, each measured with respect to their upstream sound speeds. FIGURE 4 shows schematically our simulation. A rectangular domain contains initially two constant states of gas connected by a single, left-facing shock wave of strength P_2/P_1. A unique choice of boundary values at the left- and right-hand sides of the computational box, which satisfy the Rankine–Hugoniot jump conditions, maintain the shock in a steady state. The evolution of the jet satisfying the disruption criterion is shown in the videotape Segment 4. One sees that immediately after the jet head has crossed the external pressure jump, a planar shock wave is established within the jet perpendicular to the flow direction. As a result, the postshock flow is subsonic.

The jet disruption and flaring can be understood from simple analytical arguments in two-dimensional gas dynamics. First, the pressure jump across the shock depends only upon the upstream Mach number, $P_2/P_1 = 2\gamma M_1^2(\gamma - 1)/(\gamma + 1)$. If the Mach number of the wind shock, M_{wind}, exceeds that of the jet, M_{jet}, the postshock jet

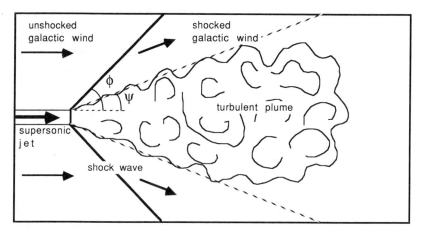

FIGURE 4. Sketch of the shock-disrupted jet, showing regions and angles referred to in the text. The jet disrupts provided the pressure jump across the internal jet shock is less than the pressure jump across the external shock, leading to the disruption criterion $V_{\text{jet}}/C_{\text{jet}} < V_{\text{wind}}/C_{\text{wind}}$, where all quantities are measured upstream of the shock waves.

pressure will be less than the postshock wind pressure. By the Bernoulli theorem, the postshock jet must further decelerate to increase its pressure to match the ambient medium. A subsonic flow expands as it decelerates, unlike a supersonic flow, hence the jet flares out. Second, the external shock near the jet must become oblique to divert the wind around the expanding subsonic tail. Referring to FIGURE 4, the oblique angle of the shock, ϕ, can be found by equating the pressure jump through the oblique shock to that of the jet shock—$\sin \phi = M_{\text{jet}}/M_{\text{wind}}$. The angle of the contact discontinuity, Ψ, separating jet and ambient material is given by

$$\cot \Psi = \frac{M_{\text{jet}}}{\sqrt{M_{\text{jet}}^2 + M_{\text{wind}}^2}} \left(\frac{(\gamma - 1)M_{\text{wind}}^2}{2(M_{\text{jet}}^2 - 1)} - 1 \right).$$

It is found that the predicted values and the measured values of ϕ and Ψ agree to within 5° for the cases run.

Whereas galactic wind shocks form at the extremities of galaxies (i.e., tens of kiloparsecs from the nucleus), the pressure wall of a transonic cooling inflow forms at radii well inside 1 kpc.[54,55] Thus, mechanism (1) previously described is the likely explanation for jets that are observed to flare as they enter the galactic halo. Similar morphological transitions are observed, however, in much smaller scale jets within central dominant cluster galaxies containing cooling inflows (e.g., M87 jet). Sumi and Smarr[56] showed on the basis of a linear stability analysis that jets emerging through cooling inflow atmospheres rapidly evolve into a region of unstable propagation. They argued that this could account for this new class of "smothered" radio sources. Soker and Sarazin[54] performed numerical simulations of a supersonic jet passing through an idealized pressure wall in order to see if the jet is disrupted. Their results were inconclusive because they were treating the ambient fluid as a static pressure field, thus ruling out mass entrainment. Recently, Jun-Hui Zhao and I repeated Soker and Sarazin's calculations, allowing the jet and ambient gases to interact self-consistently. We found that jet disruption occurred whenever the pressure at the sonic point exceeded the ram pressure of the incident jet. This criterion is entirely equivalent to that previously derived for the shock disruption mechanism described earlier.

5. THE EFFECTS OF ADDITIONAL PHYSICS

5.1. Magnetic Fields

Because radio jets radiate by virtue of their embedded magnetic fields, it is tempting to inquire what dynamical effects magnetic fields may have of radio source morphology and stability. Simulations of axisymmetric jets with strong toroidal magnetic fields have been performed[57,58] that show that the gas that has been processed through the terminal shock does not flow laterally into a cocoon, but rather is magnetically collimated into a growing "nose cone" at the tip of the jet. As a consequence, these jets do not possess cocoons *per se,* and thus resemble certain lobeless quasar jets (e.g., 3C 273). Clearly, these nose cones cannot grow to arbitrary length before the classic fire-hose instability of plasma columns sets in. Three-dimensional MHD simulations are needed to assess the generality and stability of nose cones.

In addition, Kössl, Müller, and Hillebrandt[59] have simulated axisymmetric jets with equipartition poloidal and helical fields. They find that a purely poloidal field is capable of decollimating the jet, whereas a purely toroidal field increases confinement, but introduces pinching waves. These oscillations can be stabilized by an additional poloidal field without destroying collimation.

5.2. Radiative Cooling

Radiative cooling due to collisional excitation and recombination can be important in the energy budget of protostellar jets. Blondin, Fryxell, and Königl[60] have recently performed simulations of axisymmetric, supersonic jets with cooling included. They find that the structure of such jets is similar to the Standard Model with several important differences. First, they find that a cold, dense shell condenses out of the shocked gas at the head of the jet, provided that the cooling length in either the jet or ambient gas is of order or less than the jet radius. This shell is dynamically unstable and fragments into clumps. These clumps may correspond to Herbig–Haro objects, which in some cases appear like clumpy bow shocks (e.g., HH 34S). Radiative cooling also decreases the size and pressure of the cocoon, resulting in narrower jets with fewer internal shocks. For very high cooling rates, the condensed material collects at the head of the jet, forming a high-density "nose cone" similar to that found in the magnetically confined jets described earlier. As before, the three-dimensional stability of this structure is uncertain.

6. THE STANDARD MODEL REVISITED: QUESTIONS FOR FURTHER STUDY

6.1. Status of Three-dimensional Simulations

In Section 2 the basic fluid dynamical structures of axisymmetric, supersonic jets were introduced (the Standard Model; cf. FIG. 1). In Section 3, the constraints of axisymmetry were removed by studying slab jets, and a new effect was discovered: jet flapping. Jet flapping was found to so significantly alter the axisymmetric flow structures (cf. FIG. 3) that one is prompted to ask whether the Standard Model has any relevance to fully three-dimensional jets. The slab jet simulations, coupled with linear stability analyses, yield a partial answer to this question. Namely, that axisymmetric flow structures can be expected for jet lengths of less than 6–7 growth lengths of the dominant nonaxisymmetric instability; that is, about 10 MR_{jet} for the fundamental kink or helical mode. This length can be quite large for sufficiently high jet Mach numbers. Thus, assuming a Mach number of 25 for the HH34 jet,[33] one predicts a straight jet of 100 jet diameters—approximately what is observed.

Ultimately, the question of axial stability can only be addressed by three-dimensional simulations. Several attempts have been made in this direction, but with inconclusive results. Arnold and Arnett[23] simulated the Standard Model in three-dimensional Cartesian geometry. Their simulations reproduced all the features of the Standard Model, and in addition, revealed azimuthal cocoon vortices. They speculated that the combination of azimuthal and poloidal vortices known from two-dimensional simulations could produce vortex tubes that spiral around the jet, and that these could be responsible for the emission filaments observed in some radio lobes.[3] Unfortunately, their simulations assumed a quadrantal symmetry about the jet axis, thus jet bending or flapping was precluded.

Williams and Gull[61] modeled the formation of multiple hot spots by propagating a jet out a certain distance and then suddenly changing its input direction. Simulations were performed in the upper half-plane of the jet, removing a key symmetry plane found in the work of Arnold and Arnett. Although formally three-dimensional, the number of computational zones employed normal to the base plane was so low (two zones across the jet radius) that jet flapping was numerically suppressed. Nevertheless, they were able to reproduce an essential feature of double hot spots in many powerful radio galaxies; namely, that one component is extremely compact (the one the jet is pointing to), and the other component is more diffuse. They proposed on the basis of this calculation that the compact component is the site of current jet impact, and the the diffuse component is a "splatter spot."

6.2. Future Research Topics

My own research group is developing two separate three-dimensional hydro codes for jet simulations, one based on finite-difference methods,[62] and one based on smoothed particle hydrodynamics.[63] In the context of the Standard Model, one would like to investigate such questions as: (1) the structure and stability of the working surface, (2) vortex shedding and the evolution of eddies in the cocoon, (3) strength and persistence of internal shock waves in the supersonic beam, (4) dynamical mechanism for multiple hot spots, and (5) origin of the "dentist drill" effect.[44] In particular, one would like to determine whether and to what extent jet flapping occurs in three dimensions. In the context of radio galaxy/cluster IGM interactions, one needs to study the problem of a jet in a crosswind. An attempt at this problem by Williams and Gull[64] has yielded encouraging first results. Simulations need to be performed, however, that address the internal shock structures that form inside a bent jet, as well as the role (or lack thereof) of the active galaxy's interstellar medium and wake region on jet structure and stability.

In the realm of protostellar jets, one needs to perform three-dimensional simulations of dense jets to assess the effects of beam instabilities and thermal instabilities at the working surface. It has been proposed[65] that breakup of the dense shell at the working surface of a radiatively cooling jet into dense knots may produce the clusters of Herbig–Haro objects that are observed to be aligned with the jet axis.[66] Three-dimensional simulations will be required to determine the properties and the kinematics of these knots for comparison with observations.

ACKNOWLEDGMENTS

I wish to thank my collaborators, both past and present, for their contributions to the work reviewed here: Jack Burns, David Clarke, Phil Hardee, Martin Sulkanen, Larry Smarr, Mike Smith, Paul Wiita, Karl-Heinz Winkler, and Jun-Hui Zhao.

REFERENCES

1. BRIDLE, A. H. & R. A. PERLEY. 1984. Ann. Rev. Astron. Astrophys. **22:** 319.
2. PERLEY, R. A., A. H. BRIDLE & A. G. WILLIS. 1984. Astrophys. J., Suppl. Ser. **54:** 291.
3. PERLEY, R. A., J. W. DREHER & J. COWAN. 1984. Astrophys. J., Lett. **285:** L35.
4. CLARKE, D. A., J. O. BURNS, A. H. BRIDLE, R. A. PERLEY & M. L. NORMAN. 1990. Submitted for publication in Astrophys. J.
5. STROM, K. M., S. E. STROM & J. STOCKE. 1983. Astrophys. J., Lett. **271:** L23.
6. MUNDT, R. & J. W. FRIED. 1983. Astrophys. J., Lett. **274:** L83.
7. MARGON, B. 1982. Science **215:** 247.
8. POETZEL, R., R. MUNDT & T. P. RAY. 1989. Astron. Astrophys. **224:** L13.
9. SHU, F. H., F. G. ADAMS & S. LIZANO. 1987. Annu. Rev. Astron. Astrophys. **25:** 23.

10. REES, M. J. 1982. *In* Extragalactic Radio Sources, D. S. Heeschen and C. M. Wade, Eds. Reidel. Dordrecht, The Netherlands.
11. MUNDT, R. 1985. *In* Protostars and Planets II, D. C. Black and M. S. Mathews, Eds.: 414. University of Arizona Press. Tucson.
12. BEGELMAN, M. C., M. J. REES & R. D. BLANDFORD. 1984. Rev. Mod. Phys. **56:** 255.
13. NORMAN, M. L. & K.-H. A. WINKLER. 1986. *In* Astrophysical Radiation Hydrodynamics, Vol. C188, K.-H. A. Winkler and M. L. Norman, Eds.: 187. Reidel. Dordrecht, The Netherlands.
14. WOODWARD, P. R. 1986. *In* Astrophysical Radiation Hydrodynamics, Vol. C188, K.-H. A. Winkler and M. L. Norman, Eds.: 245. Reidel. Dordrecht, The Netherlands.
15. NORMAN, M. L., L. L. SMARR, K.-H. WINKLER & M. D. SMITH. 1982. Astron. Astrophys. **113:** 285.
16. NORMAN, M. L., L. L. SMARR & K.-H. A. WINKLER. 1983. *In* Astrophysical Jets, A. Ferrari and A. G. Pacholczyk, Eds.: 227. Reidel. Dordrecht, The Netherlands.
17. NORMAN, M. L., K.-H. A. WINKLER & L. SMARR. 1984. *In* Physics of Energy Transport in Extragalactic Radio Sources, NRAO Workshop No. 9, A. H. Bridle and J. A. Eilek, Eds.: 150.
18. WILSON, M. J. & P. A. G. SCHEUER. 1983. Mon. Not. R. Astron. Soc. **205:** 449.
19. SMARR, L. L., M. L. NORMAN & K.-H. A. WINKLER. 1984. Physica D **12:** 83.
20. SMITH, M. D., M. L. NORMAN, K.-H. A. WINKLER & L. SMARR. 1985. Mon. Not. R. Astron. Soc. **214:** 67.
21. WILLIAMS, A. G. 1985. Ph.D. Dissertation, University of Cambridge, Cambridge, England.
22. ARNOLD, C. N. 1985. Ph.D. Dissertation, University of Michigan, Ann Arbor.
23. ARNOLD, C. N. & W. D. ARNETT. 1986. Astrophys. J., Lett. **305:** L57.
24. WOODWARD, P. R., D. H. PORTER, M. ONDRECHEN, J. PEDELTY, K.-H. WINKLER, J. CHALMERS, S. W. HODSON & N. J. ZABUSKY. 1987. *In* Science and Engineering on Cray Supercomputers, J. E. Aldag, Ed.: 557. Cray Research. Minneapolis.
25. LIND, K. 1988. Ph.D. Dissertation, California Institute of Technology, Pasadena.
26. KÖSSL, D. & E. MÜLLER. 1988. Astron. Astrophys. **206:** 204.
27. BLANDFORD, R. D. & M. J. REES. 1974. Mon. Not. R. Astron. Soc. **169:** 395.
28. BEGELMAN, M. C. & D. CIOFFI. 1989. Astrophys. J., Lett. **345:** L21.
29. NORMAN, M. L., L. L. SMARR & K.-H. A. WINKLER. 1985. *In* Numerical Astrophysics, J. Centrella, J. LeBlanc, and R. Bowers, Eds.: 88–123. Jones & Bartlett. Portola Valley, Mass.
30. PERLEY, R. A., A. G. WILLIS & J. SCOTT. 1979. Nature **281:** 437–442.
31. O'DEA, C. & F. OWEN. 1986. Astrophys. J. **301:** 841.
32. SHONE, D. & I. W. A. BROWNE. 1986. Mon. Not. R. Astron. Soc. **222:** 365.
33. BÜRKE, T., R. MUNDT & T. P. RAY. 1988. Astron. Astrophys. **200:** 99.
34. FERRARI, A., E. TRUSSONI & L. ZANINETTI. 1983. Astron. Astrophys. **125:** 179–186.
35. HARDEE, P. E. 1984. Astrophys. J. **287:** 523.
36. PAYNE, D. G. & H. COHN. 1985. Astrophys. J. **291:** 655.
37. NORMAN, M. L. & P. E. HARDEE. 1988. Astrophys. J. **334:** 80–94.
38. ADAMSON, T. C. & J. A. NICHOLLS. 1959. J. Aerosp. Sci. **26:** 16.
39. HARDEE, P. E. & M. L. NORMAN. 1988. Astrophys. J. **334:** 70–79.
40. ―――. 1989. Astrophys. J. **342:** 680–685.
41. FERRARI, A., S. MASSAGLIA & E. TRUSSONI. 1982. Mon. Not. R. Astron. Soc. **198:** 1065.
42. ZHAO, J.-H., J. O. BURNS, P. E. HARDEE & M. L. NORMAN. 1990. Submitted for publication in Astrophys. J.
43. HARDEE, P. E. & M. L. NORMAN. 1990. Astrophys. J. **365:** 134–158.
44. SCHEUER, P. A. G. 1982. *In* Extragalactic Radio Sources, D. S. Heeschen and C. M. Wade, Eds. Reidel. Dordrecht, The Netherlands.
45. WALKER, R. C., J. M. BENSON & S. C. UNWIN. 1987. Astrophys. J. **316:** 546.
46. HARDEE, P. E. 1982. Astrophys. J. **257:** 509.
47. ―――. 1986. Astrophys. J. **303:** 111.
48. ―――. 1987. Astrophys. J. **313:** 607.
49. ―――. 1987. Astrophys. J. **318:** 78.

50. HARDEE, P. E., M. L. NORMAN, T. KOUPELIS & D. A. CLARKE. 1990. Accepted for publication in the May 1991 Astrophys. J.
51. EILEK, J. A., J. O. BURNS, C. P. O'DEA & F. N. OWEN. 1984. Astrophys. J. **278:** 37–50.
52. O'DONOGHUE, E. & F. N. OWEN. 1986. In Radio Continuum Processes in Clusters of Galaxies, C. P. O'Dea and D. Uson, Eds.: 155–160. National Radio Astronomy Observatory. Green Bank, W. Va.
53. SOKER, N. & C. SARAZIN. 1989. Astrophys. J. **327:** 66.
54. NORMAN, M. L., J. O. BURNS & M. E. SULKANEN. 1988. Nature **335:** 146–149.
55. WHITE, R. E. III & C. SARAZIN. 1987. Astrophys. J. **318:** 612.
56. SUMI, D. M. & L. SMARR. 1984. In Physics of Energy Transport in Extragalactic Radio Sources, NRAO Workshop No. 9, A. H. Bridle and J. A. Eilek, Eds.: 160.
57. CLARKE, D. A., M. L. NORMAN & J. O. BURNS. 1986. Astrophys. J., Lett. **311:** L63.
58. LIND, K. R., D. G. PAYNE, D. L. MEIER & R. D. BLANDFORD. 1989. Astrophys. J. **344:** 89.
59. KÖSSL, D., E. MÜLLER & W. HILLEBRANDT. 1990. Astron. Astrophys. In press.
60. BLONDIN, J. M., B. FRYXELL & A. KÖNIGL. 1990. Astrophys. J. **360:** 370–386.
61. WILLIAMS, A. G. & S. F. GULL. 1985. Nature **313:** 34.
62. CLARKE, D. A., J. M. STONE & M. L. NORMAN. 1990. Bull. Am. Astron. Soc. **22**(2).
63. BALSARA, D. S. & M. L. NORMAN. 1989. Bull. Am. Astron. Soc. **21**(4):1093.
64. WILLIAMS, A. G. & S. F. GULL. 1984. Nature **310:** 33.
65. BLONDIN, J. M., A. KÖNIGL & B. A. FRYXELL. 1989. Astrophys. J. **337:** L37.
66. SCHWARTZ, R. D. 1983. Annu. Rev. Astron. Astrophys. **21:** 217.

Simulation and Visualization of Compressible Convection in Two and Three Dimensions[a]

DAVID H. PORTER, PAUL R. WOODWARD, W. YANG,
AND QI MEI

Department of Astronomy
and
Supercomputer Institute
University of Minnesota
Minneapolis, Minnesota 55415

INTRODUCTION

The combination of a supercomputer, accurate and efficient numerical methods, and hardware and software devoted to the analysis and visualization of supercomputer simulations constitutes a numerical laboratory for performing a wide variety of experiments in fluid dynamics. Over the last 4 years we have built up such a numerical laboratory at the University of Minnesota. Here we will present some of the results that this laboratory has produced. These results are best seen in the laboratory itself as color movies shown at high resolution. In this paper we can unfortunately give only a glimpse of the vast quantity of data from our experiments. Readers who wish to see the movies associated with the simulations described here should send a blank VHS, SVHS, 8 mm, or ¾-inch U-Matic video cassette and self-addressed, stamped envelope to the authors, and a video copy of the movies will be returned to them in the appropriate format.

Our numerical laboratory at Minnesota is built from several components. First, of course, is the Cray-2 supercomputer of the Minnesota Supercomputer Institute, which has performed all the simulations discussed in this article. Second is the Piecewise-Parabolic Method (PPM), a gas dynamics simulation code derived from the early MUSCL code of Woodward[1,2] and developed at Livermore by Woodward and Colella.[3–6] Third is a high-speed graphics system that stores data from the simulations, analyzes and archives it, and that generates and displays high-resolution

[a]The work reported here was performed at the University of Minnesota Supercomputer Institute. The University of Minnesota has given a great deal of support to this work. In addition to research startup grants, the University has made substantial grants of time on its Cray-2 computer and has supported purchases of graphics-oriented equipment from Silicon Graphics and Gould. Equipment donations from Sun Microsystems and Gould were also essential in supporting this work. Visualization hardware was also purchased under an equipment grant from the Air Force Office of Scientific Research, Grant AFOSR-86-0239. More recently, work reported here has been supported by the Office of Energy Research of the Department of Energy, under Contract DE-FG02-87ER25035, and by the National Science Foundation, under Grant AST-8611404. The present upgrade of our visualization equipment is supported by these two agencies and by a donation of equipment from Imprimis Technology.

color movies showing the flow dynamics. This graphics system was originally a duplicate of the one developed by Karl-Heinz Winkler's group at the Los Alamos National Laboratory.[7] We are now in the process of upgrading and replacing most of the equipment in order to allow the analysis and visualization of three-dimensional fluid flow simulations. The design principles for our new graphics system are set out in reference 8.

NUMERICAL TECHNIQUES

In the simulations discussed here we have used the PPM code to solve the Euler equations of inviscid, compressible fluid flow. For comparison of these solutions with corresponding solutions of the Navier–Stokes equations of viscous, compressible fluid flow with heat conduction, we have used the PPMNS code, which has been derived from PPM.

The PPM code was developed at Livermore to solve fluid flow problems involving strong shocks. It is based upon ideas of Bram van Leer for extending the classic method of Godunov[9] to a higher formal order of accuracy. The foundation of PPM is a piecewise-parabolic interpolation scheme that generates a parabola to describe the internal structure of a computational cell, or zone, of the grid. A parabola is generated for any desired variable based upon knowledge of the zone-averaged values of that variable. The parabola is constrained to be monotone increasing or decreasing and to attain values that lie within the limits set by the minimum and maximum zone-averaged values in the zone and in its two neighbors.

An important feature of PPM is an algorithm for detecting contact discontinuities and for redefining the parabolas near such discontinuities. In these regions the parabolas are made more consistent with a sharp jump in the variable value and less consistent with an assumption of smooth flow. This discontinuity detection and steepening permits the PPM code to move discontinuities great distances across the grid without spreading them out over more than one or two zones. This feature of PPM is most useful in multifluid hydrodynamics problems and in treating contact discontinuities with slip, which can arise from Mach shock intersections.

Our implementation of PPM is directionally split. That is, we take into account derivatives in the *x*-direction in one sweep and follow that by a similar sweep to account for *y*-derivatives. The sweeps are symmetrized to give second-order formal accuracy.[10] Each directional sweep is further split into a Lagrangian step followed by a remapping to the original Eulerian grid. The remap step is essentially determined by the piecewise-parabolic interpolation scheme and the need to conserve mass, momentum, and total energy exactly. This exact conservation, which is obeyed in the Lagrangian step as well, guarantees that shocks whose internal structure does not change as they move through the grid will have the correct speeds and satisfy the correct jump conditions. For a discussion of problems that can arise when internal shock structures are not steady, the reader is referred references 4 and 6.

In recent years we have applied the PPM code to fluid flow problems that do not involve strong shocks. Although these are not the kind of flows that PPM was designed to compute, we have found that the method is extremely effective in this flow regime. Of course, for these problems we have removed from PPM features such as nonlinear approximations to the solution of Riemann's problem and complicated formulations of dissipation terms required only in pathological cases that do not arise in these gentler flows. The resulting simplified code runs 40 percent faster than the more robust version and is adequate for problems involving shock pressure jumps up to about a factor of 10.

A prime advantage of the PPM code is its elaborate formulation of numerical dissipation. This dissipation has been constructed with great care, and with considerable computational cost, in order to balance the sharpness of shock structures against the numerical noise that they inevitably generate in certain pathological cases.[4,6] In the convection simulations to be presented here, where only weak shocks are encountered, one might well ask if this numerical dissipation of the PPM code is still appropriate. We believe that indeed it is. Just as in the case of shocks, our concern in these convection problems is to generate as little dissipation as possible on length scales of several grid zones while nevertheless dissipating the kinetic energy of smaller scale modes into heat as strongly as possible. This behavior will allow the code to resolve as many modes as possible in an "inviscid" turbulent cascade while still turning the energy of the smallest scale modes into heat before it can be unphysically transmitted to the larger scales.

In order to quantify the behavior of the numerical dissipation in PPM, we have observed its effects empirically on a number of different generic flow structures.[11,12] For sinusoidal shear waves introduced at a 45° angle to the mesh, these studies indicate the following formula for the effective kinematic viscosity, v_{eff}, of PPM:

$$\frac{v_{eff}}{cL} = 0.462 \left(M + \frac{1}{4}\right)^{3.15} \left(\frac{L}{\Delta x}\right)^{-3}. \tag{1}$$

Here c is the sound speed in the gas and M is the Mach number, the ratio of the amplitude of the velocity variation to c. The wavelength of the sinusoidal velocity variation is $\sqrt{2}L$. To obtain Eq. (1), cases were examined in which $2L$, the periodic length in either coordinate direction, varied from 4 to 128 zones and the Mach number varied from 0.01 to 0.7. The observed behavior is fit to about 20 percent accuracy for $L/\Delta x > 16$ and the fit is still useful even for the smallest mesh resolutions tested.

We conclude from our studies, which are presented briefly below, that PPM is an efficient and accurate tool for the simulation of time-dependent fluid flow not only for problems involving strong shocks but also for compressible flow problems in which only weak shocks or even no shocks at all are generated. The reason for this is the presence of discontinuities even in shockless flows. The discontinuities may mark boundaries between different fluids, as in the slip surface simulations below, or they may mark internal viscous and thermal layers that are generated during the course of the flow, as with our convection experiments. In either case, PPM keeps these discontinuities sharp while generating a minimum of numerical noise in the flow around them. PPM is thus a means of achieving a high effective Reynolds number in an Euler flow simulation, as Eq. (1) suggests.

SIMULATIONS OF COMPRESSIBLE CONVECTION IN TWO DIMENSIONS

Our work on compressible convection was inspired by the earlier efforts of Hurlburt, Toomre, and Massaguer.[13] These authors solved the Navier–Stokes equations governing the flow of a stratified viscous fluid with heat conduction in two dimensions that was confined between two impenetrable, friction-free plates. At the bottom plate, heat was introduced at a constant rate, while at the top plate a constant temperature was maintained. Hurlburt *et al.* began their numerical experiments with an unstable equilibrium state and let the flow evolve until a steady flow was reached. Applying periodic boundary conditions in the horizontal direction, they obtained steady convection in a single convection cell with a width of four times the depth of their unstable fluid layer. This result was surprising because the density in their initial state varied by a factor of 21 from the top to the bottom of the layer. Standard heuristic theories of convection in such stratefied layers suggest the development of many smaller eddies. Our work on this subject began at the suggestion of Toomre and Hurlburt that we apply our PPM code to this problem in order to investigate a higher effective Reynolds number regime than is attainable with Navier–Stokes simulations.

The purpose of this study of compressible convection is to obtain a greater understanding of the process of heat and material transport in the outer layers of stars like the sun. In these stars thermal conduction carries heat effectively within the deep interior where the gas is highly ionized. The mechanism for termal conduction is the diffusion of radiation. This process is much more effective than atomic diffusion in transporting heat, because the mean free path of the radiation is so much larger. Closer to the stellar surface, the gas becomes cooler, and recombination of hydrogen and helium atoms makes the gas much more opaque. In this region the heat flux cannot be maintained stably by radiation diffusion, and therefore convection results.

The convection zone in a star like the sun spans many density and pressure scale heights, and is characterized by an extremely small thermal conductivity and an even smaller viscosity. To directly simulate these conditions with the Navier–Stokes equations on a computer, even in only two space dimensions, is at present an impossible task. Hence in our PPM simulations we have used the relatively small effective viscosity of the code parametrized in Eq. (1) and have added a very small thermal conduction term. The thermal conduction is treated explicitly in a straightforward manner in the code.[11] In the simulations presented here it is extremely small in a dimensionless sense, with a coefficient of heat conduction of either 10^{-4} or 10^{-5}.

Our initial efforts in our study of compressible convection have been reported in Porter and Woodward,[14] and later work is presented in Woodward *et al.*[15] Also, the last of a series of three hour-long video lectures[16] focuses on convection and presents detailed color animations of several of our convection simulations. The dramatic dependence of the convection of a stratified fluid upon the effective viscosity can be seen in FIGURE 1, where three PPM simulations of the same problem are compared at several different times. These simulations differ only by their grid resolutions, and hence by their effective viscosities. All three runs began from the same initial

(a)

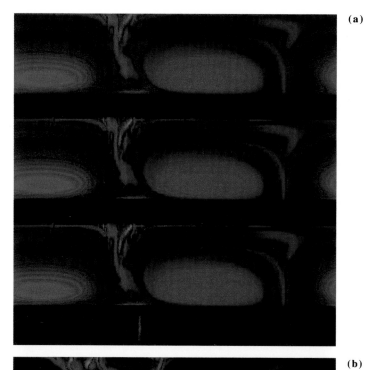

(b)

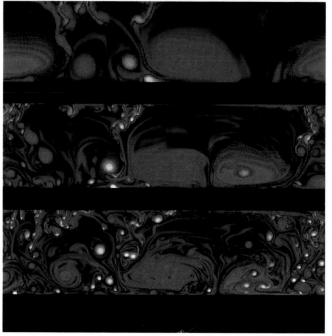

FIGURE 1 (a) and (b)

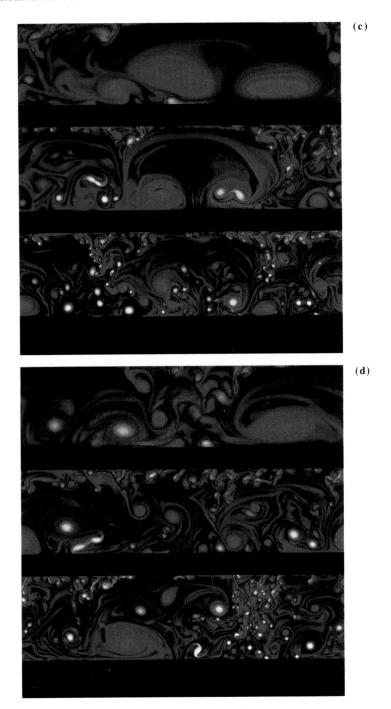

(c)

(d)

FIGURE 1 (c) and (d)

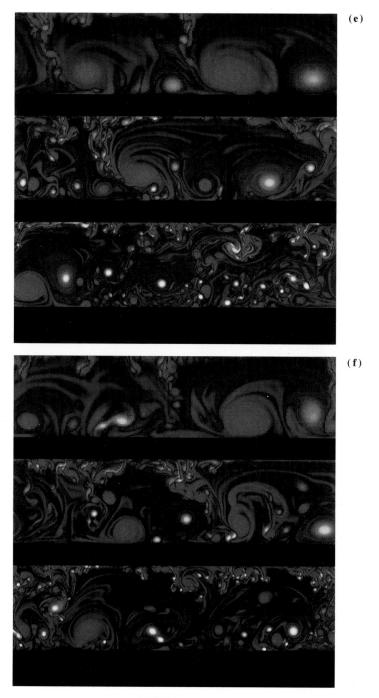

FIGURE 1 (e) and (f)

condition. This initial condition was obtained by simulating the convection on a still coarser grid of only 128 × 32 square computational zones.

To obtain the initial condition for the runs in FIGURE 1, we began with an initial condition like that used by Hurlburt *et al.* On a grid of 128 × 32 square zones (Hurlburt *et al.* used 160 × 40 zones) we set up an unstable equilibrium polytrope in which heat is carried by thermal conduction from the bottom wall to the top. We use a gamma-law equation of state, with $\gamma = 5/3$, to describe an ideal monatomic gas (presumably hydrogen). To give a constant heat flux throughout the layer we let the temperature, and hence the internal energy in the gas, depend linearly on the depth, z. Taking the density linearly proportional to the depth as well gives us a polytrope of index 1. We take a constant gravitational acceleration, g, in the layer, and we choose our units so that its value is unity. Hydrostatic equilibrium then demands that $dp/dz = g\rho$, so that the pressure increases as the square of the depth. If we measure the depth z downward from the surface, where the density vanishes, then our layer is chosen to extend from $z = 0.05$ to $z = 1.05$. By choosing the constant of proportionality between the density ρ and the depth z to be unity, we fix our unit of mass. The unit of length is the thickness of the layer, and the unit of time is fixed by setting g to unity. The free-fall time from the top to the bottom of the layer is then $\sqrt{2}$. The turnover time for the largest eddies also turns out to be near unity in all flows presented here.

In our simulations we apply periodic boundary conditions in the horizontal direction. At the bottom wall we introduce heat at a constant rate by thermal conduction. Therefore we keep the vertical derivative of the gas internal energy constant at the bottom wall. At the top wall we keep the temperature of the gas constant. The initial unstable equilibrium state chosen for our simulation on the 128 × 32 grid was consistent with these boundary conditions. In order to observe its convective instability, it was given a small white noise perturbation in the velocity, with an amplitude of about 1 percent of the sound speed at the bottom of the layer. High-resolution simulations of the onset and development of convective instability from such an initial state are presented in references 14–16 and in 11.

The three simulations in FIGURE 1 begin with a flow obtained by evolving the initial state previously described on a grid of 128 × 32 square zones for an extremely long time (thousands of time units), until it reaches a statistically steady state. This state is quite similar to the truly steady flow obtained with the Navier–Stokes equations by Hurlburt *et al.* from a similar starting point. The effects of reduced effective viscosity are observed by interpolating the flow from the 128 × 32 grid to a

---

FIGURE 1. A comparison of PPM simulations of compressible convection in a two-dimensional layer carried out at three different grid resolutions, and hence with different effective viscosities. The vorticity distributions in these flows are shown at times 0, 100, 200, 300, 400, and 500. All three simulations begin from the same initial condition, which is the nearly steady convection flow obtained on a grid of 128 × 32 zones. From the top to the bottom of the figure, the grids used are: 256 × 64, 512 × 128, and 1024 × 256 square computational zones. The density variation from the top to the bottom in these flows is about a factor of 40. A coefficient of thermal conduction of 10^{-5} was used, so that in the best resolved simulation the thermal boundary layer at the upper boundary is only three to four zones in thickness. As the effective viscosity is reduced there is a transition from a flow with one major convection cell to flows with two such cells and finally to a chaotic, completely unsteady flow in which no major, persistent convection cells can be identified.

256 × 64 grid (top), a 512 × 128 grid (middle), and a 1024 × 256 grid (bottom) and continuing the calculation for an additional 500 time units. The snapshots of the three flows in FIGURE 1 show that as the effective viscosity is reduced there is a transition first to an organized convection with two large convection cells each spanning the entire thickness of the layer, and later to a chaotic convection in which no large, persistent convection cells can be identified. This last transition to truly chaotic convection in the highest resolution simulation requires about 200 time units. This need for both high grid resolution and simulation for hundreds of dynamic times makes these compressible convection runs challenging for even the fastest computers.

A detailed analysis of the results of these convection simulations for a variety of interesting parameter values can be found in Porter and Woodward.[11] By observing the flows in detail using high-speed color graphics in our numerical laboratory we can identify the mechanisms that give rise to the spectrum of eddy sizes observed in the final, statistically steady flow. A thin thermal boundary layer forms near the top boundary, so that the heat transported up from the bottom wall by the convection can be carried out of the system by means of thermal conduction. At this top boundary the gas is cooled and is then compressed by the surrounding gas. The cool compressed gas is then less bouyant and sinks. This gas forms thin plumes as it descends, for this is very much like a Rayleigh–Taylor instability. The descending plumes generate small vorticies at their tips and also along their lengths by means of the Kelvin–Helmholtz shear instability. This, rather than any sort of breakup of larger eddies, is the primary mechanism for the formation of small vortices in the flow.

Large eddies are generated and maintained mainly by the merger of eddies of the same sense of spin. Eddy mergers are common, so that a modest-sized eddy is more likely to disappear as a result of merger than as a result of numerical viscous dissipation. This sort of vortex interaction has been studied in detail in the incompressible regime by Zabusky and his coworkers.[17] We note that the process of eddy merger can result in the shearing apart of small eddies and in the production of thin filaments of vorticity. These filaments of vorticity tend to be aligned with the fluid streamlines. We have analyzed the velocity power spectra from the simulations shown in FIGURE 1 in considerable detail. The results are presented in Porter and Woodward. For the simulation on the finest grid these spectra show a clear knee at a wavelength of about 12 zone widths. For wavelengths shorter than this a much steeper power law dissipation sets in that is caused by the effective numerical viscosity of PPM.

The images of the vorticity in the gas that appear in FIGURE 1 give a good representation of the spectrum of eddy sizes in the flow, and, when animated, they also show the velocity field as a whole quite well. They do not, however, indicate the significant acoustic energy that is present. Like the vorticity, acoustic energy is generated on all scales. In some simulations, a large standing acoustic wave with a wavelength equal to the width of the layer is present that shapes the behavior of the flow on large length and time scales. In all our simulations, shock waves develop that feed acoustic modes on the small scales.

We find that for our flows with small thermal conductivity and small effective viscosity the eddies tend to spin at substantial fractions of the local speed of sound.

Indeed, some of our eddies spin so rapidly that the centrifugal forces cause the gas density near their centers to be reduced by up to an order of magnitude compared to the surrounding region. Large eddies in this highly stratified fluid can turn subsonically near the bottom of the layer, where the sound speed is highest, and supersonically near the top. Shocks develop naturally under such circumstances, and their interactions with the many eddies in the flow are fascinating to observe. FIGURE 2 shows the divergence of the velocity in a simulation of the onset of convective

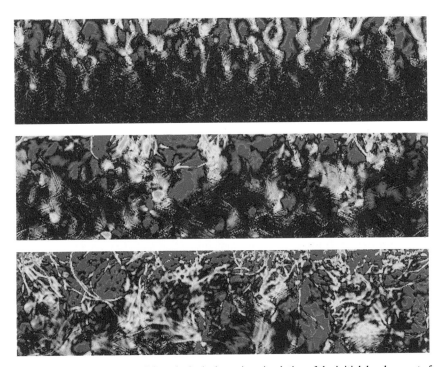

FIGURE 2. The divergence of the velocity is shown in a simulation of the initial development of the convective instability of an unstable polytropic equilibrium state (polytropic index 1). The 3 images correspond to times 8, 20, and 32. A grid of 1024×256 square zones was used, and the initial state was as described in the text. The coefficient of thermal conductivity in this simulation was 10^{-5}. The steady build up of acoustic energy in the flow and the development of many shock waves throughout the flow is most clearly seen in this display.

instability that was carried out at high resolution. The many shocks show up quite clearly, as well as the general correlation between compression and descending fluid motions. After seeing our results for compressible convection flows containing shocks, Cattaneo, Hurlburt, and Toomre[18] verified that on sufficiently fine computational grids their Lax–Wendroff approach based on the Navier–Stokes equations also produced shock phenomena despite the much higher effective viscosities that characterize their simulations to date.

We have tested the accuracy of our two-dimensional convection simulations in two ways. The comparison shown in FIGURE 1, of course, amounts to such a test. Indeed, we wish we could afford to carry out such a simulation on a 2048 × 512 grid for a period of 500 dynamic times. Unfortunately, our Cray-2 is not fast enough to permit such an experiment. Hence, we can only hope that on our finest grid we have captured the correct qualitative convective behavior for a gas of very low viscosity. We have performed other, more affordable tests. First, we have extensively investigated the accuracy with which our PPM code treats the process of the merger of two eddies.[12] For this study we have built a Navier–Stokes version of PPM. This PPMNS code allows us to compute viscous eddy mergers that are not marred by any numerical effects. The Navier–Stokes runs are marred only by the relatively small Reynolds numbers that can be achieved with grids of up to 1024 × 1024 zones in size. The PPMNS code is described in Woodward.[19]

SIMULATIONS OF COMPRESSIBLE EDDY MERGER

In FIGURE 3 are shown vorticity distributions from two simulations of the merger of two identical eddies, each initially spinning at about Mach 1/2 and with the initial configuration shown. (For a more detailed discussion of these and other related simulations, see Yang *et al.*[12]) The eddies are confined in a box with reflecting,

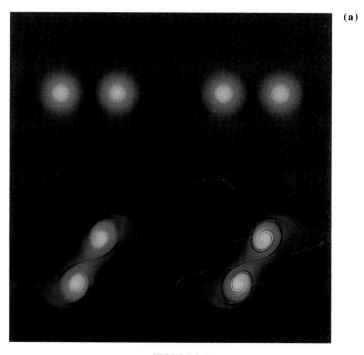

(a)

FIGURE 3 (a)

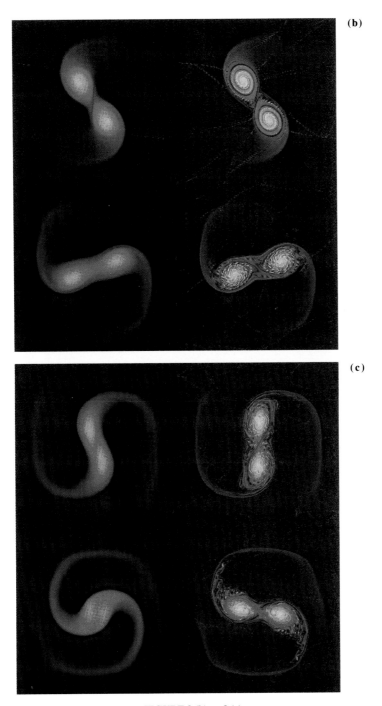

FIGURE 3 (b) and (c)

(d)

(e)

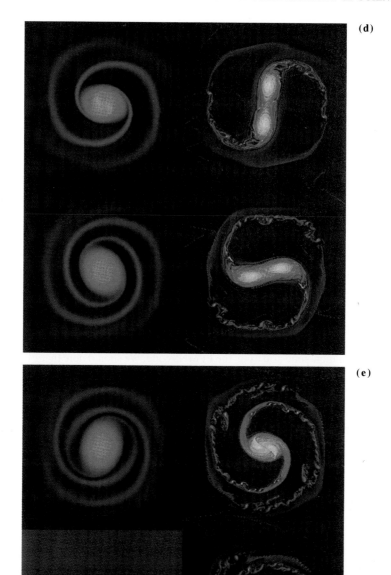

FIGURE 3 (d) and (e)

friction-free walls. The confining box was introduced so that this problem could be easily set up by others and could easily be attacked using spectral methods. The box does not have a dramatic influence on the eddy merger, although its effects are visible in the shape of the outer strands of vorticity near the end of the simulations.

These eddies were chosen to be representative of those we find in our convection simulations. Therefore, they are initially isothermal. They are initially given Gaussian distributions of vorticity, with a standard deviation of one length unit and with an amplitude that yields a maximum velocity of about Mach 1/2. The initial eddy separation is 7.1 length units, and the confining box is 20.48 length units on a side. Initially, the velocity fields of the two eddies are simply added together, and the pressure gradient is set to balance the centrifugal forces that the resulting velocity field implies. In the far field, the density and the sound speed are set to unity for this ideal monatomic gas ($\gamma = 5/3$). The pressure is then determined from instantaneous force balance, while the density is determined by the initial isothermal constraint.

The initial eddy interaction is intense. It results in the generation of very thin layers of strong shear as well as two principal shock waves. These shocks eventually reflect from the walls of the box, and other, weaker shocks develop there because the normal component of the gas velocity does not initially vanish at the walls. The two simulations shown in FIGURE 3 have been computed on very fine grids of 1024 × 1024 square zones. FIGURE 3 compares results obtained using the Navier–Stokes equations with the PPMNS code and using the Euler equations with the PPM code. The Navier–Stokes simulation treats an ideal monatomic gas with the smallest coefficients of heat conduction and viscosity that are consistent with an accurate simulation on this million-zone grid using the PPMNS code. We should note that PPMNS is extremely accurate. It can accurately calculate the propagation and dissipation of a sinusoidal sound wave with a wavelength as small as just three zones. The Euler simulation treats an ideal monatomic gas in which both heat conduction and viscosity are assumed negligible. For very short wavelength disturbances, PPM falsifies the behavior implied by Euler's equations. Truncation errors as well as explicitly added dissipative features of PPM act to produce this falsification. Nevertheless, PPM gives an extremely accurate approximation to the so-called "weak solution" of Euler's equations.

FIGURE 3. Two simulations of the merger of two identical eddies are displayed here at time intervals of 25 starting at time 0 and ending at time 200. The eddies spin initially at about Mach 1/2, and they are confined in a box with reflecting, friction-free walls. The eddies are initially placed close together with pressure forces balancing the instantaneous centrifugal forces implied by their joint velocity field. Densities are determined initially by setting the temperature constant throughout and by setting the density and sound speed to unity in the far field. At the left a Navier–Stokes simulation for an ideal monatomic gas is shown. On the right an Euler simulation using the PPM code is displayed. Both simulations were carried out on grids of 1024 × 1024 square zones. The PPM run shows the development of Kelvin–Helmholtz instabilities along the very thin, strong shear layers that arise near the outset of both of these runs. Because of these instabilities, a well-defined solution to this flow problem in the limit of vanishingly small viscosity is elusive, and this comparison cannot demonstrate convergence of the PPM simulation to such a desired flow. This study does, however, indicate the futility, with the resources provided by present supercomputers, of nearing any such limit by solving the Navier–Stokes equations numerically.

The comparison of these two simulations of compressible eddy merger yields many fascinating insights into the behavior of both means of approximating the flow that we really wish to simulate, namely a Navier–Stokes flow with coefficients of heat conduction and viscosity many orders of magnitude smaller than those used in the PPMNS run. To put this into perspective, if the gas involved were air at sea level and at normal room temperature, the box in our PPMNS simulation would measure roughly 0.01 cm on a side. On coarser grids than those shown in FIGURE 3, PPMNS produces obvious falsifications of the flow we are really looking for. The diffusion of vorticity in the eddy cores due to viscosity of the gas is striking on these coarser grids. It is even clearly visible on the million-zone grid. There is very little hint of any shear flow instability in the PPMNS runs, despite the very thin and strong shear layers that are generated near the outset of the simulation. Presumably, viscous dissipation is simply too great in these simulations for any instability to develop.

In contrast to the PPMNS simulations, the PPM runs appear to converge rapidly to an overall rotation rate of the merging eddy cores about one another. On the finer grids the development of shear instabilities is enhanced. Of course, each grid refinement unveils instabilities on still smaller length scales, but between the 256×256 grid and the 512×512 grid there is a marked increase in unstable behavior, because the large-scale shear of the differentially rotating flow tends to oppose growth of the largest modes. We have examined results of the 1024×1024 PPM run in great detail and see no obvious connection between known numerical falsifications (other than numerical viscosity, of course) of PPM and the excitation of the particular unstable modes observed in this run. It would be most encouraging if we could demonstrate similar behavior in a PPMNS run on a finer grid, but the Cray-2 is too slow to make such a simulation practical. Roughly similar behavior to the 1024×1024 PPMNS run is observed with a 128×128 PPM run (see FIG. 4), and the two codes run at the same speed on the Cray-2 to within about 20 percent. Hence we may conclude that our approach in using PPM to investigate the high Reynolds number regime for compressible convection is certainly cost effective. It is also clear, however, that a wealth of detail is hidden from view in even the best resolved of the runs in FIGURE 1.

SIMULATIONS OF COMPRESSIBLE CONVECTION IN THREE DIMENSIONS

The convection simulations described earlier have been motivated by a desire to understand the dynamics of the solar convection zone. These simulations, however, are of course very idealized. The solar convection zone does not have a floor and a ceiling, nor is it flat, two-dimensional, and periodic. In fact, the sun rotates and has imbedded magnetic fields. The equation of state of the solar gas and the behavior of its transport coefficients with temperature is also more complex than assumed in our work. Clearly, there are many improvements that we can make so that our simulations can be made more applicable to convection in the sun. Of the many possible improvements we have chosen to concentrate upon the dimensionality of the simulations, because we feel that the restriction of the flow to two dimensions is the most drastic simplification we have made.

(a)

(b)

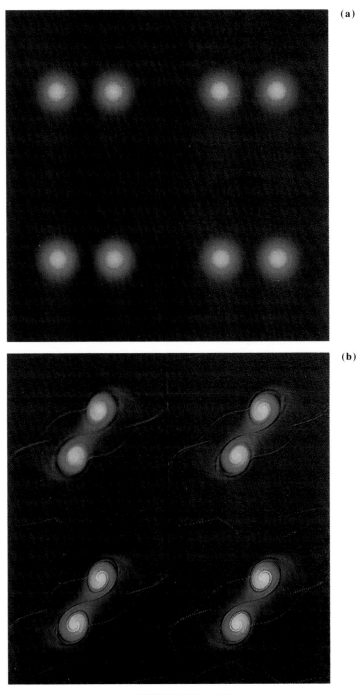

FIGURE 4 (a) and (b)

(c)

(d)

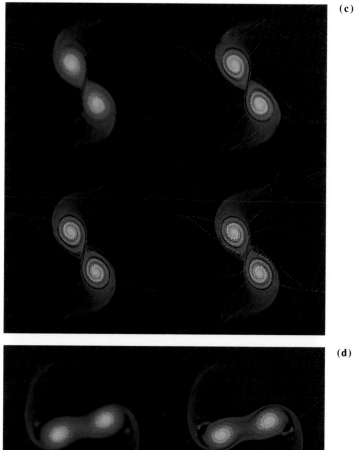

FIGURE 4 (c) and (d)

(e)

(f)

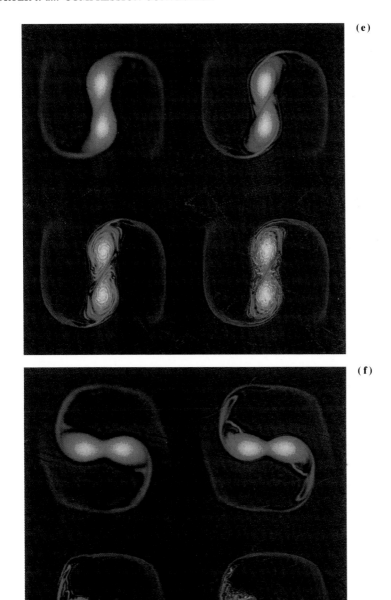

FIGURE 4 (e) and (f)

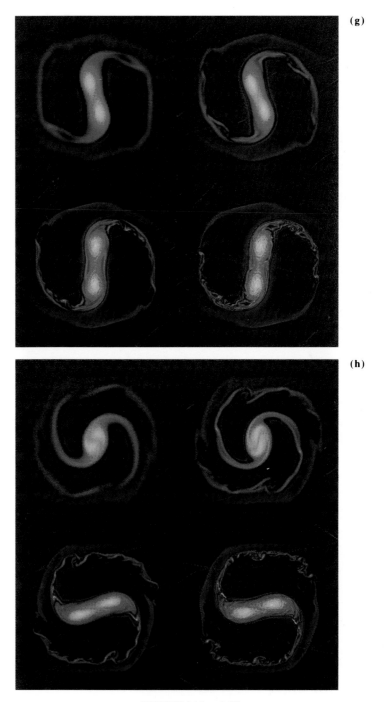

FIGURE 4 (g) and (h)

(i)

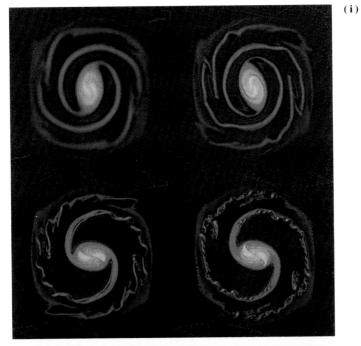

(j)

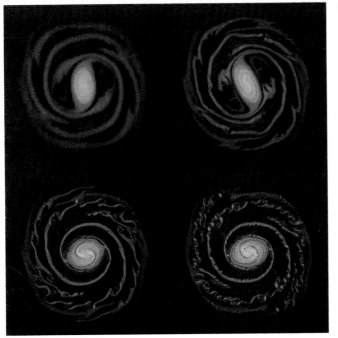

FIGURE 4 (i) and (j)

(k)

(l)

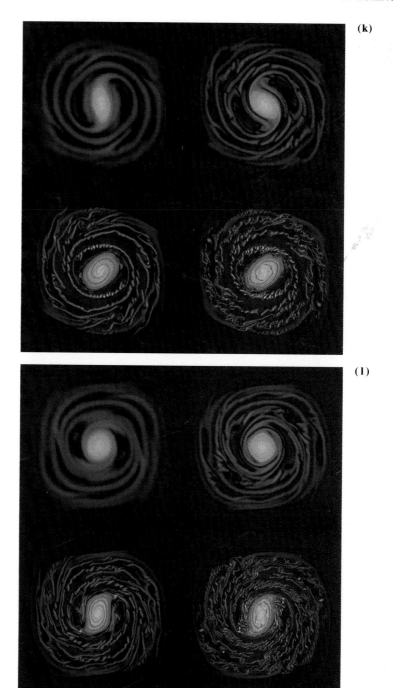

FIGURE 4 (k) and (l)

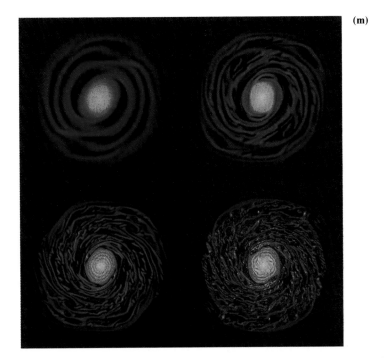

(m)

FIGURE 4. Results of 4 PPM simulations of the same problem shown in FIGURE 3 are displayed here at the same time in the evolution of this problem. Grids of 128, 256, 512, and 1024 zones on a side were used to obtain these results. These should be compared with the Navier–Stokes results in FIGURE 3. The 128 × 128 grid seems to produce results most closely resembling those of the million-zone Navier–Stokes run (at roughly 500 times less cost). The simulations are shown at time intervals of 25, beginning at time 25 and ending with time 300.

One might well ask why we did not simply begin our study of compressible convection with three-dimensional simulations. In fact, we have had a functioning three-dimensional version of our PPM code for about three years. There is, however, a good deal more to three-dimensional simulation than simply developing a three-dimensional fluid dynamics code. In fact, there is little point in performing such a simulation unless one has the means of analyzing and visualizing the data to be computed. It was not until quite recently that we felt it wise to undertake a program of three-dimensional fluid flow simulation. Very large amounts of supercomputer time are needed for these runs, truly massive amounts of data are produced, and these data must be visualized and analyzed if the dynamics is to be understood. These considerations are described in some detail in Woodward and Winkler,[8] but it will suffice here to give a single example.

In FIGURE 5 we show five snapshots of a three-dimensional compressible convection simulation that was performed on the Cray-2 at the University of Minnesota last spring. This run illustrates both the problems and the benefits that come along with three-dimensional fluid flow simulation. The simulation was designed to investigate the stability of the two-dimensional convection flows that we

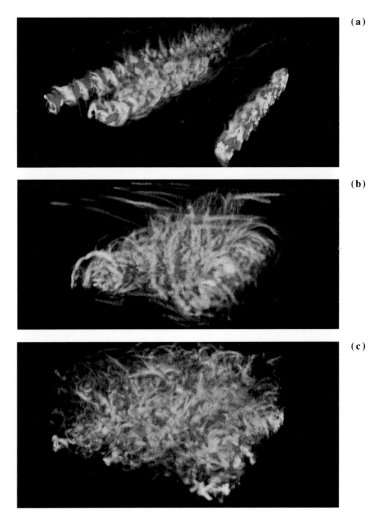

FIGURE 5. A statistically steady two-dimensional convection flow on a grid of 200×100 square zones was used to generate the initial condition for this three-dimensional simulation. The parameters of this run were similar to those shown in FIGURES 1 and 2, except that the coefficient of thermal conduction was set to 10^{-4}. The two-dimensional flow was replicated in the third dimension and perturbed with a white noise additional isotropic velocity with an amplitude of 0.01, or about 1 percent of the sound speed at the bottom of the layer. These displays have been generated by a volume-rendering technique that makes only those regions with large magnitudes of the vorticity visible. These regions are assigned colors, ranging from black to white, according to the value of the local helicity, the dot product of the velocity with the vorticity. A kink instability of the original line vortices can be seen that results in a total rearrangement of the flow. These snapshots have been taken at times 10, 20, and 30.

have been discussing in the fourth section. We began with an initial condition like that used in the flow shown in FIGURE 2, or like that used to generate the initial conditions for the flows in FIGURE 1. This was a polytropic unstable equilibrium state with a polytropic index of 1. The coefficient of thermal conduction was chosen to be 10^{-4}. Our layer was only twice as wide as it was deep for this simulation, and we used a relatively coarse grid of 200×100 square zones. We ran this simulation in two dimensions for a very long time until it reached a statistically steady state, which consisted of a single large convection cell. We then replicated this flow in the third dimension and applied a small, isotropic, white noise velocity perturbation to the flow, with amplitude 0.01 (about 1 percent of the sound speed near the bottom of the layer). This small velocity perturbation broke the perfect symmetry of the flow, and as a result its evolution from this point had to be computed in three dimensions.

The three-dimensional computation required 250 CPU hours on the Cray-2, and this run went on continuously for 10 days. At 500 times during the course of the run data were dumped to disk. Compressed data dumps from which any image might be generated for this run are each 50 Mbyte in size. Had we saved all these data, the run would have generated 25 Gbyte for later visualization and analysis. We are now building a dedicated fast disk farm capable of holding this amount of data from a simulation, but last spring it was not possible to contemplate saving these data. Instead, at the outset we chose 10 movies to be rendered, and only these images were saved at all 500 time intervals. Only 50 of the compressed data dumps were saved and 6 full data dumps from which the run might be restarted. The 6 full dumps make up 1.2 Gbyte, the 50 compressed dumps make up 2.5 Gbyte, and the 10 movies of 500 frames each, with each frame 3/4 Mbyte, made up 3.75 Gbyte. This is a total of 7.45 Gbyte of information that we saved. Every 4 hours our partition of the shared disks of the Minnesota Supercomputer Center had to be cleared in order for the run to proceed. Thus, this simulation was unusually labor intensive.

The images shown in FIGURE 5 were generated by a volume rendering technique that we developed for the visualization of three-dimensional continuous flows.[20] This technique was inspired by a Pixar demonstration, and it extends the methods used in that demonstration to full perspective rendering from any viewpoint, including viewpoints within the flow itself, and it accomplishes this entirely within the context of a fully vectorizable and parallelizable Fortran code. The development of this rendering technique was crucial in making this three-dimensional simulation practical.

In FIGURE 5 the convection flow is shown by associating an opacity with each zone of the grid. This opacity is chosen to be proportional to the magnitude of the vorticity squared. Opaque regions of the flow hide the flow behind them from view, and at the same time they emit light in proportion to their opacity so that they become visible. The light these regions emit is associated with a second, independent flow variable, the helicity. The helicity is the dot product of the flow velocity vector and the vorticity vector. In FIGURE 5 regions of large vorticity are visible, while other regions are transparent. Hence, we see the vortex tubes from the original two-dimensional flow most clearly in the early snapshots of the three-dimensional simulation. These vortex tubes appear to be striped like barber poles with regions of opposite helicity. Gas with one sign of helicity travels along a vortex tube in one direction, describing a helix as it flows, while gas of the opposite helicity proceeds along the vortex tube in the opposite direction. The banded appearance of the vortex tubes indicates the presence of a kink instability of these tubes that can be more easily perceived in the high-resolution animation of these images.

By the end of the run in FIGURE 5, the originally two-dimensional flow has been completely disrupted. This shows that, not surprisingly, the two-dimensional convection flows that we have been studying in great detail for several years are probably all violently unstable when they are permitted the additional degree of freedom represented by the third dimension. This three-dimensional example also indicates what difficult work we will have in store for us in untangling and understanding the far more complicated dynamics that these far more realistic systems exhibit. It will require powerful computers, powerful numerical techniques, and powerful visualization and analysis of the simulations in order to develop an understanding in this exciting new area of research.

REFERENCES

1. VAN LEER, B. 1979. J. Comput. Phys. **32:** 101.
2. VAN LEER, B. & P. R. WOODWARD. 1979. The MUSCL code for compressible flow: Philosophy and results. *In* Proceedings TICOM Conference, March.
3. WOODWARD, P. R. & P. COLELLA. 1981. Lect. Notes Phys. **141:** 434.
4. ———. 1984. J. Comput. Phys. **54:** 115.
5. COLELLA, P. & P. R. WOODWARD. 1984. J. Comput. Phys. **54:** 174.
6. WOODWARD, P. R. 1986. *In* Astrophysical Radiation Hydrodynamics, K.-H. Winkler and M. L. Norman, Eds. Reidel. Dordrecht, The Netherlands.
7. WINKLER, K.-H., J. W. CHALMERS, S. W. HODSON, P. R. WOODWARD & N. J. ZABUSKY. 1987. A numerical laboratory. Phys. Today.
8. WOODWARD, P. R. & K.-H. WINKLER. 1989. Simulation and visualization of fluid flow in a numerical laboratory. Interdisciplinary Sci. Rev. In press.
9. GODUNOV, S. K. 1959. Nat. Sb. **47:** 271.
10. STRANG, G. 1968. SIAM J. Numer. Anal. **5:** 506.
11. PORTER, D. H. & P. R. WOODWARD. 1989. Simulations of compressible convection with the piecewise-parabolic method (PPM). In preparation.
12. YANG, W., P. R. WOODWARD & D. H. PORTER. 1989. Simulation of eddy merger in the compressible regime. In preparation.
13. HURLBURT, N. E., J. TOOMRE & J. M. MASSAGUER. 1984. Astrophys. J. **282:** 557.
14. PORTER, D. H. & P. R. WOODWARD. 1988. Simulations of compressible convection using the piecewise parabolic method (PPM). *In* High Speed Computing, Scientific Applications and Algorithm Design, R. B. Wilhelmson, Ed. University of Illinois Press. Urbana.
15. WOODWARD, P. R., D. H. PORTER, M. ONDRECHEN, J. A. PEDELTY, K.-H. WINKLER, J. W. CHALMERS, S. W. HODSON & N. J. ZABUSKY. 1987. Simulations of unstable fluid flow using the piecewise-parabolic method (PPM). *In* Science and Engineering on Cray Supercomputers. Cray Research. Minneapolis, Minn.
16. WOODWARD, P. R. 1988. Supercomputer simulations in astrophysics. A series of three hour-long video programs produced by the University of Minnesota Department of Independent Study for broadcast over the National Technological University closed circuit network. Broadcast Nov. 8, 1988.
17. MELANDER, M. V., N. J. ZABUSKY & J. C. MCWILLIAMS. 1988. J. Fluid Mech. **195:** 303.
18. CATTANEO, F., N. E. HURLBURT & J. TOOMRE. 1989. Supersonic compressible convection. *In* Annual Research Report of the John von Neumann National Supercomputer Center. Fiscal Year 1988: 35.
19. WOODWARD, P. R. 1989. Numerical techniques for gas dynamical problems in astrophysics. *In* Computational Astrophysics, P. R. Woodward, Ed. Academic Press. New York.
20. PORTER, D. H. 1989. Perspective volume rendering. In preparation.

Nonlinear Astrophysical Fluid Dynamics: The Video

MICHAEL L. NORMAN

National Center for Supercomputing Applications
and
Department of Astronomy
University of Illinois at Urbana-Champaign
Urbana, Illinois 61801

INTRODUCTION

A videotape has been assembled containing animations shown by speakers at the Nonlinear Astrophysical Fluid Dynamics Conference. This videotape forms a useful supplement to the conference proceedings. The contents and availability of this videotape are described herein.

Nonlinear astrophysical fluid dynamics phenomena are inherently nonstationary and complex, as the contents of this videotape attest. The best way to display time-dependent phenomena, whether produced through observations or simulations, is in the form of animations. This videotape contains a number of such animations from current astrophysical research contained in these conference proceedings. Most animations are from numerical simulations, which, as you will see, are beginning to approach the spatial and temporal complexity of the real world. A few, however, are from the real world. So turn on your TV set and enjoy!

CONTENTS

I. **Solar System**
  (1) E. Spiegel   Solar granulation film
  (2) P. R. Woodward   Compressible convection
  (3) NASA/JPL   *Voyager* sequence of Jupiter's red spot
  (4) T. Dowling   Simulation of Jupiter's red spot
II. **Galactic**
  (6) D. Arnett   Supernova instabilities
  (7) K. Prendergast   Multifluid interstellar medium
III. **Extragalactic**
  (8) M. Norman   Astrophysical jets

AVAILABILITY

The videotape is available from NCSA for the cost of materials ($6 for ½'' tapes; $12.50 for ¾'' tapes) and shipping. When ordering, please use the address below and specify the number of tapes and the format (VHS, Beta, U-Matic) desired. Please write to:

> Scientific Visualization and Media Services
> National Center for Supercomputing Applications
> 152 Computing Applications Building
> 605 E. Springfield Avenue
> Champaign, IL 61821

ACKNOWLEDGMENTS

I would like to thank Cordelia Baron of the Scientific Visualization and Media Services group for her creative input and hard work on this project. Thanks also go to Robert Buchler, Conference Organizer and Proceedings Editor for his patience, and to the University of Florida, Gainesville, for providing financial assistance. Finally, I would like to thank the video contributors.

RELATED READINGS

Other Florida Workshops in Nonlinear Astronomy published by the New York Academy of Sciences.

1. BUCHLER, J. R. & H. EICHHORN, EDS. 1987. Chaotic Phenomena in Astrophysics. N.Y. Acad. Sci. **497.**
2. BUCHLER, J. R., J. R. IPSER & C. A. WILLIAMS, EDS. 1988. Integrability in Dynamical Systems. N.Y. Acad. Sci. **536.**
3. BUCHLER, J. R. & S. T. GOTTESMAN, EDS. 1989. Galactic Models. N.Y. Acad. Sci. **596.**
4. BUCHLER, J. R., S. DETWEILER & J. R. IPSER. 1991. Nonlinear Problems in Relativity and Cosmology. N.Y. Acad. Sci. In press.

Epilog: Some Quotations from the Conference[a]

Nature has a problem, too. ED SPIEGEL

Make things violent, just for fun. PAUL WOODWARD

You have only seen a snapshot of the interstellar medium. KEVIN PRENDERGAST

Ignore gravity, because it's evil. JOHN LATTANZIO

There are no free lunches, but there are quite a few good expensive ones. KEVIN PRENDERGAST

How about crud? DAVE ARNETT

Convection, convection. Bad Problem! GEORGE BOWEN

There is never a sensible order to anything. You learn more from a shuffled deck. STIRLING COLGATE

We are forced to do it correctly. JIM TRURAN

One or two years. DIMITRI MIHALAS

---

[a] A short selection from the pithy remarks inspired by the heightened atmosphere of the conference.—ED.

Index of Contributors